ALL ABOUT GROWING

FRUITS & BERRIES

1 THE HOME FRUIT GARDEN

PAGE 5. Design fruit and berry plants into your landscape through careful planning and the use of dwarf varieties.

2 CLIMATE

PAGE 13. Use the zone map and the accompanying climate descriptions to select the varieties of fruit best suited to your area.

3 PLANTING AND CARE

PAGE 23. Information on planting, feeding, watering, and mulching your fruits and berries. Illustrations and solutions to pest and disease problems.

4 PRUNING AND TRAINING

PAGE 35. Complete illustrations and descriptions of pruning and training techniques for all your fruit and berry plants.

5 FRUIT IN CONTAINERS

PAGE 51. Planting fruits and berries in containers allows you to move them around and to grow tender varieties that would not otherwise survive your climate.

6 ENCYCLOPEDIA

PAGE 57. Special care requirements and color photographs are followed by flavor, color, and time of harvest information for hundreds of varieties.

ALL ABOUT GROWING

FRUITS & BERRIES

Created and designed by
the editorial staff of ORTHO Books

Edited by
Barbara Ferguson

Graphic design by
Barbara Ziller and John Williams

Illustrations by
Ron Hildebrand

Ortho
Books

Publisher
Robert L. Iacopi

Editorial Director
Min S. Yee

Managing Editors
Anne Coolman
Michael D. Smith

Photographic Director
Alan Copeland

System Managers
Chris Banks
Mark Zielinski

Senior Editor
Sally W. Smith

Editors
Jim Beley
Susan Lammers
Deni Stein

Production Manager
Laurie Sheldon

Photographers
Laurie A. Black
Richard A. Christman
Michael D. McKinley

System Assistant
William Yusavage

Photo Editors
Anne Dickson-Pederson
Pam Peirce

Production Editors
Alice E. Mace
Kate O'Keeffe

Production Assistant
Darcie S. Furlan

National Sales Manager
Garry P. Wellman

Sales Assistant
Susan B. Boyle

Operations/Distribution
William T. Pletcher

Operations Assistant
Donna M. White

Administrative Assistant
Georgiann Wright

Address all inquiries to:
Ortho Books
Chevron Chemical Company
Consumer Products Division
575 Market Street
San Francisco, CA 94105

Chevron Chemical Company
575 Market Street, San Francisco, CA 94105

ACKNOWLEDGEMENTS

Color Separations:
Color Tech Corp., Redwood City, CA

Contributing Editor:
Will Kirkman

Contributing Illustrator:
Leavitt Dudley

Consultants:
James Beutel
Davis, CA

Fay Paquette
Camarillo, CA

Auburn University, AL
Harry J. Amling
Joseph D. Norton
Henry P. Orr
Raymond L. Self

University of California
Claron O. Hesse
H.P. Olmo
Perley Payne, Extension Service
Robert G. Platt

Horticultural Research Institute
of Ontario, Canada
Alec Hutchinson

University of Oregon, Mid-Columbia
Experiment Station
W. M. Mellenthin

Northwestern Washington Research
and Extension Unit
Robert A. Norton

New York Experiment Station at Geneva
Robert D. Way

Photographers:
Names of photographers in alphabetical
order are followed by page numbers on
which their work appears. R = right,
C = center, L = left, T = top, and
B = bottom.

William C. Aplin: 46T

John Blaustein: 35

Clyde Childress: 59TL

Josephine Coatsworth: 4, 8T, 9T, 8-9,
10T, 10B, 10R, 33, 84L, 97L

Derek Fell: 60R, 61L, 61R, 61C, 61B,
63C, 65L, 65TR, 66C, 67L, 69TR,
69BR, 70L, 70C, 71TC, 73C, 81L,
81TC, 81R, 84R, 85C, 85B, 86L,
86C, 86R, 87B, 88L, 89R, 90C, 92R,
93R, 94L, 95L, 95TC, 95C, 95BC,
96C, 96R, 99L, 99R, 102L, 105C, 105R

Pamela Harper: 5, 83BR, 91R, 92L

Susan Lammers: 107

Michael Lamotte: Front cover

Michael Landis: 7B, 16-17, 23, 52L,
73R, 80L, 91L, 98BC

John Lund: Back cover BR, 18-19,
58R, 64L, 64R

Michael McKinley: 50, 51

Ortho Photo Collection: Back cover
TL, TR, BL, 6B, 6-7, 14B, 15, 16, 34,
52R, 53T, 53B, 55, 56, 58L, 59C,
59TR, 60L, 62L, 62TC, 62R, 62C,
62B, 63L, 63R, 63B, 65R, 66L, 66R,
67C, 67R, 68L, 68R, 69L, 69C,
69CR, 70TC, 70R, 70C, 70B, 71L,
71TR, 71C, 71R, 71B, 72L, 72TC,
72BC, 72R, 74R, 75R, 76C, 79L,
79C, 80T, 81C, 81BC, 82L, 82C, 83L,
83TR, 85L, 85R, 87L, 87TR, 88-89,
93L, 94R, 95R, 96L, 96TR, 97R, 98L,
98TC, 98R, 98C, 99TC, 99C, 99B,
100L, 100TC, 100R, 100C, 101L,
101C, 101R, 101B, 102R, 102C,
102B, 103L, 103R, 105TL, 105L,
106L, 106R

William Reasons: 14T, 46B, 57, 59B,
63TC, 90L, 90R

Stark Brothers Nursery, Louisiana,
MO: 74L, 75L, 76L, 76R, 77L, 77R,
78C, 78R

Wolf von dem Bussche: 73L

Dave Wilson Nursery, Hughson, CA:
78L, 82TR, 82BR

Photo Research:
Colour Library International (USA)
Limited: 12, 13, 22

Photo Stylist:
Sara Slavin: Front cover, 4, 6, 7, 8, 9,
10, 33

Typography:
CBM Type, Sunnyvale, CA

*Front cover: Fresh fruits and berries
provide a rewarding harvest. For
more information see "Encyclopedia
of Fruits and Berries" on pages
57-106.*

*Back cover: Apples are described on
pages 58-64, pears on pages 80-82,
grapes on pages 97-102, and cherries
on pages 66-68.*

*Title page: 'White Imperial' currants
surpass 'White Grape' currants in
quality, although both varieties are
quite similar. See page 96.*

Orthene ® is a registered trademark
of the Chevron Chemical Company.

Let your garden do more for you.
When you select dwarf varieties
and plan carefully, you can grow
a wide range of fruits and berries
no matter how small your space.

THE HOME FRUIT GARDEN

Strictly speaking, a fruit is the seed-bearing portion of any plant; but the term commonly suggests the delicious edible fruits that grace our tables —apples, pears, peaches, plums, and tangy sweet berries of all kinds. In this book you'll find all the most popular fruits and berries suitable for growing by the home gardener. The book also presents many of the best varieties for all parts of the country with descriptions of their important characteristics and their cultural requirements.

Many gardeners share the misconception that producing a good fruit crop requires the knowledge and skill of the expert orchardist. In fact, extensive maintenance and culture are necessary only for maximum commercial production. The home gardener can get by with less complicated spraying, feeding, and pruning programs. In addition, we now have improved varieties with better fruiting, disease resistance, and tolerance of special soil and climate conditions. Your chances of growing good fruit are greater today than ever before. If you learn about your growing conditions, choose the varieties that fit them, and give your plants the attention and care we outline in this book, you are bound to grow good fruit.

The climate section beginning on page 13 gives valuable information on your climate and how well the fruits and berries will perform there. Information on planting, general care, and maintenance begins on page 23; and special needs for each plant can be found under the entries in the

Modern dwarf fruit trees make it possible to tuck many fruit trees into a small space. This vegetable garden bears fruit without losing space for vegetables.

"Encyclopedia of Fruits and Berries," beginning on page 57.

FRUITS IN THE LANDSCAPE

Gardeners who grow fruits and berries at home will tell you that the taste of fresh-picked fruit and berries is more than enough reason to grow them; but fruit plants also enhance the landscape —even in a small yard or garden.

Not long ago a gardener with an average-size lot had to be content with very little in the way of fruit—perhaps a single apple tree in the center of the lawn and a grapevine growing over an outbuilding or arbor. But as the average lot has grown even smaller, modern horticulture has met the challenge. Fruit trees are now available in a range of sizes that permit using them almost anywhere—in the smallest yards and gardens—spotted about the yard, as borders and hedges, as groundcovers, or even as small shade trees. Modern dwarfing techniques and simplified methods of training allow you to grow as many as a dozen fruit trees in the same small garden and still have plenty of sunny space available for vegetables or flowers.

Because of these possibilities, this book emphasizes small-space gardening. This first chapter focuses on dwarf fruit trees. Planting fruit trees and caring for them is explained on pages 23–33. On pages 35–49 we offer two techniques for keeping fruit plants compact and productive: pruning and training. Another effective means of keeping fruit plants to size is to grow them in containers; on pages 51–55 you'll find information on healthy and productive container culture. The "Encyclopedia of Fruits and Berries," beginning on page 57, includes many dwarf varieties well suited to your garden.

DWARF TREES

The key to good fruit in the small home garden lies in the effective use of dwarf fruit trees. They provide not only good fruit but also attractive shape, foliage, and flowers.

The development of dwarf trees can be readily appreciated when you consider what would happen to your gardening space if you planted a standard-size apple tree, which can easily reach anywhere from 20 to 40 feet in height, with a spread of 30 to 40 feet. The tree would take up a considerable amount of space and block out so much sun that the ground beneath it would be too dark to grow other plants. Compare this with the dwarf apple varieties that can be held to a height of about 10 feet, with a similar spread. Standard apricots, peaches, and plums can grow to 30 feet tall with a spread of 30 feet; sweet oranges to 20 to 30 feet tall with a spread of 15 to 20 feet; and pears to a towering 45 feet with a spread as wide as 30 feet. Most dwarf varieties of these trees can be kept to a height of about 10 feet—in certain cases, even less.

Genetic Dwarfs

Dwarf trees are produced in nature or through horticultural practices. Natural dwarfs are called *genetic dwarfs*. Among apples, the most common genetic dwarfs are "spur" apples, so named because on a given amount of wood they produce more fruiting spurs than ordinary apple trees. Fruit production can have a dwarfing effect, because it uses energy that would otherwise go into the growth of the tree. The heavier crops of spur apples mean slower tree growth.

These natural mutations occur with many varieties of apple trees; for example, there are several spur varieties of the popular 'Red Delicious' apple available to gardeners under such names as 'Redspur' and 'Starkrimson'. Spur apples grow more slowly but eventually reach about three-quarters normal size. This may be as much as 15 to 30 feet

tall with a similar spread—still a large tree for many home gardens. However, all genetic dwarfs can be made even smaller by grafting them onto dwarfing rootstocks.

An example of a genetic dwarf apple that is not a spur type is the variety sold as 'Garden Delicious'. It bears fruit resembling the popular 'Golden Delicious', but grows to only 6 or 8 feet; 3 feet when grown in a container.

Apricots, sweet cherries, sour cherries, peaches, nectarines, and plums all have genetic dwarf varieties. There are several good, extremely cold-tolerant, genetic dwarf sour cherries that grow to perhaps 10 feet under ideal conditions but most often stay at 6 or 7 feet. There is a wide range of genetic dwarf peaches, all of which grow slowly to 8 or 9 feet. As these have almost no stem between the leaves, they have a typically dense or lumpy look. These dwarf peaches and the dwarf nectarine (the nectarine being a *sport*, or mutant, of the peach) are quite decorative, with large, showy flowers. They are also fairly tender, however, and tolerate little cold. There are many natural dwarf plums, but they are not the popular varieties; rather, they are crosses of the Western sand cherry, or true cherry, and the shrub plum. Most are produced commercially for very cold climates and are readily available in the North.

Man-Made Dwarfs

There are several horticultural methods of producing dwarf trees and all work by limiting the supply of nutrients to the tree to retard growth.

Pruning has a dwarfing effect because it removes some of the foliage that enables the plant to grow. Heavy summer pruning will increase dwarfing.

Top: These trees have been trained to remain small.
Left: This apple is grafted onto dwarfing rootstock.
Below: Dwarfing apples by both grafting and training.

Root pruning is an effective dwarfing procedure. In this technique, a sharp spade is sunk into the ground around the perimeter of the plant to sever some of the feeder roots.

Girdling and scoring also reduce tree growth. In girdling, a strip of bark about ¼-inch wide is removed from around the tree. In scoring, a single cut is made around the tree. Both of these methods interfere with the supply of nutrients to the top of the tree.

Training by bending or twisting branches to direct growth can also have a dwarfing effect. You'll find information about training trees on pages 46–49.

All these techniques produce only a temporary dwarfing effect and must be repeated periodically. Growing in containers can produce a permanent dwarfing by confining the roots of the tree to a small growing space; however, you must be sure that the roots do not grow out of the bottom of the container and become established in the soil below it.

The easiest and most effective way to produce permanent dwarfing is by grafting scions (tree cuttings of a desired variety) to dwarfing rootstocks (rootstocks that limit the supply of nutrients). This method offers many advantages to growers, horticulturists, and gardeners, as grafting is the only way to produce large numbers of plants in a relatively short time. It also ensures true reproduction of a desired variety, which is important because seed does not always breed true. In addition, grafted dwarf trees will remain uniformly smaller. A secondary effect of grafting is that grafted dwarfs tend to bear fruit at a younger age than standard trees, sometimes as early as the second year.

Grafted dwarfs are readily available at most nurseries and garden centers. More curious, adventuresome, and enterprising gardeners may want to try their own grafting.

Malling Rootstocks

The key to producing dwarf trees is in growth-limiting rootstocks. The most extensive research on such plants has been undertaken with apples and has resulted in the development, at the Malling Agricultural Research Station in England, of the numbered Malling rootstocks. Since the roots of these special cultivars are relatively shallow and confined in growth, they limit the mature size of any fruit-bearing variety grafted onto them. There are several.Malling rootstocks available, each having a different degree of dwarfing effect on the apple variety that is grafted onto it.

Other major fruit trees can also be dwarfed by grafting, but far less research has been done on these. As a result, fewer kinds of dwarfing rootstocks are known, and these are not always as effective as the apples. Apricots, peaches, nectarines, and plums can be dwarfed on Nanking cherry rootstock, cherries on St. Lucie cherry, and citrus on sweet orange and trifoliate rootstocks. For pears, the only reasonably satisfactory rootstock is quince.

One of the advantages of growing dwarf trees is easy care. When the distance from roots to treetop is 10 feet or less, sprays for pests and diseases are easier to apply and pruning is less difficult and time-consuming. Because the roots are shallow and less extensive, you apply less fertilizer and water and thus save time and money. Moreover, these advantages increase if you reduce size further by applying the pruning and training techniques offered in this book.

Right: Fruits and berries, trained and espaliered by Mr. Whiting, fill this narrow garden area.

Photos on these pages and the next are from David Whiting's garden. Below: A peach espalier being trained. Right: A dwarfed kumquat.

A FRUIT GARDEN

The sample plan on page 11 illustrates some of the remarkable space-saving possibilities in gardening with fruit. Notice the orientation of the garden. Fruit needs sun to set a crop, and the illustrated arrangement provides maximum exposure to the sun as it passes from east to west. If your location does not have all-day sun, at least plan to give your fruit plants southern or western exposures.

Notice also how this plan makes use of espaliered apple trees, cane berries, and grapevines to provide an attractive border and plenty of fruit, and still allows space for many other plants, if desired. The berries are trained on trellises and oriented approximately north and south. They take up little space and bear heavily. The grapes are planted mainly on the north side of the garden to give them full southern exposure to the sun, which is necessary in order to develop good sugar content in the fruit. Note that there is still room in the center of the garden for dwarf fruit trees, raised beds in which to grow vegetables or flowers, a plot of annuals, a good stand of raspberries, and more.

To show how such a garden might be carried out, we have photographed a garden developed by David Whiting of St. Helena, California, which measures only 15 by 50 feet but contains 17 fruit trees, several grape varieties, cane berries, ornamental plants, and vegetables. The training of plants plays a most important part in this garden. Instructions for making espaliers like those in the photographs are on page 47. This requires a little effort, but once established, espaliered trees can be grown closer together and will still receive good light; will have perfect air circulation and plenty of root space; and will require less attention with sprays, fertilizers, and pruning shears.

A SAMPLE GARDEN

When you landscape with fruit, you combine beauty with practicality. You may not want a garden composed almost entirely of fruit, but you may want to add a few fruit plants to enhance your surroundings.

Fruit can serve many functions. For example, apple trees make superb shade trees anywhere in the yard if you prune them to a branch high enough to allow passage underneath. A large crab apple tree or a spreading cherry will provide good shade.

Any fruit tree you like can be used as a focal point or accent in the yard or garden. The most striking trees in bloom are apples, cherries, quince, and the showier flowering peaches. Crab apples are especially effective. These hardy trees are the most widely adapted of all flowering trees and offer abundant displays of red to pink to white blossoms followed by brilliantly colored fruit. As a rule they require some winter chilling, but some varieties bloom beautifully even in the mild Pacific Coast climates. Some crab apples have fragrant blossoms; others have red to purple foliage. Citrus trees, where they can be grown, also have attractive foliage, showy fruit over a long season, and a wonderful fragrance.

Shrub fruits can also play an important role in the landscape, either as individual accents or in hedges or shrub borders. Try blueberries for their subtle colors or currants for their beautiful flower clusters and brilliant scarlet fruit.

Genetic dwarf peach trees make splendid flowering hedges, and showy-flowered dwarf or standard peaches can be trained in the same way. Espaliered apples or pears can also form attractive hedges or borders.

You can even use fruit as a ground cover. Strawberry plants are effective, especially in smaller areas, but plan to replace them every three years with new plants if you want a heavy fruit crop. For larger areas, use the low-bush blueberry.

These are just a few ideas for planting with fruit. Other possibilities depend on climate, soil, available varieties, and your personal taste. This book will give you all the information you need to enjoy the many benefits of gardening with fruits and berries.

Opposite page—Top: This espaliered apple is in its last year of training. A different variety has been grafted on top.

Bottom: The fig being trained up the wall makes use of every bit of space.

Right: Salad vegetables share space with fruit in the Whiting garden.

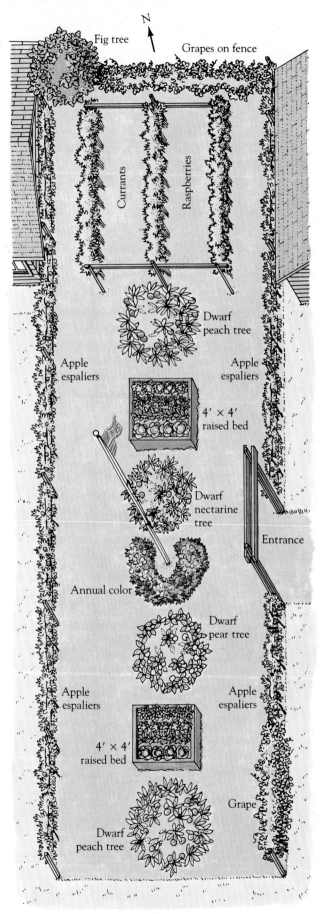

THE 15′ × 50′ SAMPLE GARDEN

N

Fig tree
Grapes on fence
Currants
Raspberries
Dwarf peach tree
Apple espaliers
Apple espaliers
4′ × 4′ raised bed
Dwarf nectarine tree
Entrance
Annual color
Dwarf pear tree
Apple espaliers
Apple espaliers
4′ × 4′ raised bed
Grape
Dwarf peach tree

*Choose the fruit and berry
varieties best suited to your climate.
This chapter includes a zone map,
complete descriptions of climates,
and lists of the fruits that
will grow in each zone.*

Most of the fruits we discuss in this book are referred to as temperate-zone fruits. The temperate zone in the Northern Hemisphere lies between about 23½° latitude and the Arctic Circle. This, however, tells little about the climatic conditions in which these plants will thrive. In the United States alone the growing season — the period between the last frost in winter and the first frost of the following winter — ranges from 3 to 12 months in length, depending on location. Climates, and all the factors that make up climate — prevailing temperatures, humidity, rainfall, cloud cover and fog, amount of sunshine, and winds — vary enormously throughout this country. Furthermore, they vary not only over large distances, but also within cities from block to block, from garden to garden, and sometimes even from place to place within the same garden. This is why in recent years gardeners have taken to thinking and planning in terms of *microclimates*.

Your own garden may have several microclimates. For example, a spot protected from the wind will have a microclimate warmer than a spot out in the open. If the sheltered spot is backed by a wall that reflects heat, that area will be even warmer. In a northern garden such a location might be ideal for helping a tree bear better fruit, whereas in a southern garden this location might produce too much heat.

The extreme diversity of climate in the United States

An early snowfall blankets late apples with snow. Modern fruit breeders have extended the range of most fruits, but it is important to select varieties adapted to your region.

and the possible variations even within individual gardens make it impossible to provide an exact guide to climatic conditions. In this chapter, however, there are some good general guidelines on climate in your part of the country and on what fruits to grow there. This information will help you choose the varieties most likely to do well in your garden. For more detailed information, be sure to check with nurseries, garden centers, agricultural extension agents, and — especially — gardening friends and neighbors who can tell you about their own successes and failures.

CLIMATE ZONES
The map on pages 20 and 21 is described below in four sections: the southern states; the midwestern and northeastern states; the western states; and, because of its wide climatic variations, California. Below is a description of the characteristics of the various zones within these four sections. The map itself is divided (by color) into zones based on the approximate length of the growing season for fruit plants and the minimum winter temperature. A color-coded chart that indicates how well the fruits and berries in this book will do in each zone, and an explanation of the symbols in each chart accompany this map.

FRUIT CLIMATES OF THE SOUTH
The southern states have such a broad range of climates that no general rules apply to the whole region. The major winter influence is a mass of continental air that moves down from the Canadian plains, sometimes reaching into the usually mild regions of the Gulf Coast and Florida.

Top: *Spring storms threaten pear blossoms.*
Bottom: *Peaches grow in all parts of the South except southern Florida.*

Severe cold spells seem to occur in the Deep South in cycles of from 10 to 20 years, with less severe spells every 4 or 5 years. A major summer influence is the warm, moist Gulf air, which reaches northward and inland up rivers and along valleys to affect areas far from the coast.

The South is a favored fruit-growing region, hospitable to a broad range of fruiting plants. However, these hot, moist summer changes do favor the spread of pests and disease. The gardener must be especially careful to check plants often, to provide good air circulation, and to spray when necessary.

When planning a fruit garden in the South, you will have to determine the variations in climates most common to your specific locality. For example, the gardener in Tennessee or Kentucky may have next to no warming influence, and cold winters are a certainty. Apples and pears will do well. Peaches should be varieties that like a long period of dormancy. Plums and sour cherries will do well, but apricots and sweet cherries may suffer in some years from fall or spring frosts.

The gardener south of Baton Rouge, Louisiana, can plant tender fruits such as citrus, but must count on damage or even loss of the more tender plants on a regular cycle of perhaps every 10 years. These plants can be replaced, and the replacements will bear fruit again in a season or two.

In borderline areas between zones you may be able to choose from a wide range of fruiting plants, but always with the understanding that damage may occur in winter and that the hot, moist summer may give you trouble with hardier plants. You'll want to choose pears that resist fireblight, a bacterial infection that is at its worst in warm spring and fall weather. The peaches you plant should be resistant to leaf spot and canker.

In recent years breeding stations in the South have developed peach varieties that grow and fruit well as far south as Florida. The pear, once limited by fireblight to a few cooking varieties, now includes a wide range of blight-resistant varieties. The territory for bunch grapes has also been extended, and now even Florida gardeners can grow some varieties.

Climate Zones of the South

The southern portion of the climate map is divided into six zones. In general, if you live in a northerly or inland portion of a zone your climate will be more severe and your season will be shorter.

Zone 1. Grow only tropical plants in this zone; for example, citrus instead of pears and peaches.

Zone 2. This zone includes the band around the Gulf Coast where citrus and subtropicals can be planted commercially. An occasional freeze may cause damage.

Zone 3. This zone has a strong tropical influence. The hardy subtropicals grow well, but some temperate-zone fruits are limited.

Zone 4. This zone is the upper section of the Cotton Belt. The short winter and long growing season call for varieties that prefer little winter cold. Disease may be a problem.

Zone 5. This zone north of the Cotton Belt is hospitable to any temperate-zone fruit, but choose varieties that tolerate cold.

Zone 6. This zone includes mountainous inland areas. The growing season is shorter and winters can be harsh. Use the hardiest temperate-zone fruits, and in winter provide protection with windbreaks and mulch.

Chart Symbols

These chart symbols are designed to help you choose the right variety of fruit for your zone and to indicate when a plant may suffer from winter cold or the lack of it.

A: All varieties. Plant any of the varieties of fruit listed in this book.

H: Hardy varieties. Choose varieties that tolerate cold. Peaches, for example, may need as many as 1,000 hours of winter temperatures under 45°F.

P: Protection in winter. In the coldest zones, heavily mulch and cover fruit plants, or grow them in containers that can be moved to shelter.

U: Uncertain crop. Some plants find the southern climate too warm in summer or too cool in winter. Blossoms may open early and be lost to frost, or the trees may freeze or succumb to disease.

R: Disease-resistant varieties only. The early southern spring in this zone encourages fireblight in some of the pear varieties.

D: Dwarf varieties. Dwarf varieties are especially good in this zone.

L: Low-chill varieties. These grow with the least winter cold. Between November 1st and February 15th they should have less than 500 hours of temperatures below 45°F. For example, only trees of this type should be planted south of Highway 84 in Mississippi.

(No symbol.) The fruit type is not recommended for this zone.

Strawberries are the only berry that can be grown in all parts of the South.

FRUIT CLIMATES
OF THE MIDWEST AND NORTHEAST

The climates of the northern region of the United States range from the harsh winters and short, cool summers of upper Michigan to the fairly mild coastal climates of New Jersey. At least some varieties of fruit will grow in each of these climate zones.

In much of the North, extreme winter cold is a serious threat to fruit. Be sure to choose the hardiest varieties. Much damage from cold depends on whether the plant is dormant: A completely dormant plant takes cold better than one that is growing. Flower buds are particularly sensitive to late frosts and freezes. Plant your garden so that cold air will flow down and away from the fruit trees when warm spring days are followed by extremely cold nights.

Another northern and midwestern threat is wind, which can dry plants out when ground water is frozen and thus unavailable to the plants. If you have no naturally protected site for planting, put in an evergreen windbreak when you put in your fruit plants. A windbreak of trees is more effective than a solid barrier such as a fence, since the wind can't flow over the top and down on the other side in a solid wave. For temporary wind protection, use snow fencing or burlap tacked to stakes.

The first strong sun can do serious damage to plants that are still leafless. You may want to protect your trees from sunburn in these situations by wrapping the trunk in burlap or by painting the trunk and main branches with white interior latex paint.

Little can be done about ice, but be careful to remove snow and icicles if they build up enough to break branches.

Above: Apples being washed.

Right: Apples are one fruit crop that can be grown in the coldest parts of the United States.

Climate Zones of the Midwest and Northeast

The midwest and northeast section of the map is divided into five zones. If you live in a northerly or inland portion of a zone, your climate will be more severe and your season shorter.

Zone 5. This zone is particularly good for growing fruit. In the mildest coastal areas, gardeners can even try a fig tree in a warm corner.

Zone 6. Fruit crops do well in this area if you protect them from the worst weather. The growing season is long, but occasional extremes can cause damage.

Zone 6a. The presence of large bodies of water in Michigan, Ohio, New York, and coastal New England makes winter less severe and summers longer. This is the case in the fruit belt of the southern portion of Michigan, where commercial orchards produce cherries, peaches, and many other fruits. You can plant any temperate-zone fruit here with good prospects for success, although you will need to

select the hardiest apricots and fairly hardy plums and peaches.

Zone 6b. This zone includes the coldest areas of New England. It corresponds to Zone 7 but has less severe winds and more snow cover. Apples are among the best tree crops. Give plants protection and use dwarf trees in containers that can be brought inside in winter.

Zone 7. This zone includes the coldest parts of the United States. The short growing season and late frosts limit your choice to the hardiest varieties of fruit plants. In the Central States winds can be intense. Windbreaks and other protection are necessary. Consider growing dwarf trees in containers that can be brought indoors for part of the year.

Chart Symbols

A: All varieties. Plant any of the varieties that are listed in this book.

H: Hardy varieties. Choose the most cold tolerant varieties of fruit.

P: Protection in winter. Dwarf apples and pears should have deep winter mulch. Where snow cover is uncertain, hold the mulch in place with fence wire or evergreen boughs. Spread the mulch 4 to 6 inches deep from the trunk all the way out to the point below the branch tips. Some plants, such as strawberries, raspberries, and grapes, can be buried in mulch after the first cold stops all growth.

U: Uncertain crops. Even where hardy fruiting plants survive, extreme cold or late frost will sometimes destroy the blossoms of young fruits. Plants with a "U" listing will do somewhat better if placed in a climate-moderating planting site, such as near a south-facing wall.

(No symbol.) The fruit type is not recommended for this zone. Plants may lack a symbol for two reasons: Either they are too tender (vulnerable to cold) for the district, or they can be replaced by superior, less tender varieties.

FRUIT CLIMATES
OF THE WEST AND CALIFORNIA

In the western states there are many prime areas for fruit growing, from the Pacific Northwest to the Salt Lake area and the Arizona desert. There are six climate zones on the map of the West, seven on the map of California. The white areas are considered too harsh for fruit crops.

Climate Zones of the West

Zone 5. Gardeners in the high desert of Arizona and New Mexico escape the worst of the summer heat and can grow apples, pears, and many other fruits. Cold, however, rules out tender plants.

Zone 6. This zone dots itself around the map. It is cold, but not nearly as cold as the related climate of Zone 7. Some influence of terrain or water lengthens the growing season and tempers the winter. Eastern Washington, for example, is prime apple country, with extensive commercial orchards, but it also produces good peaches of the high-chill Elberta type, and home gardeners can succeed with almost any temperate-zone crop. For apricots and cherries, try the hardiest varieties.

Zone 7. Related to Zone 6, but with a growing season up to 30 days shorter, this climate is still mild enough for hardy varieties of most temperate-zone fruits. There is commercial fruit production in both Colorado and Montana. Protect trees by mulching the roots heavily in winter, holding the mulch in place with wire. Plant on ground that has good drainage, and establish windbreaks.

Zone 8. Around Phoenix, Arizona, the climate corresponds roughly to the low desert regions of California, producing fine grapefruit and mandarin oranges. Temperate-zone fruits such as apples and pears will not survive here.

Zone 8a. The milder, warmer parts of southern Washington and western Oregon produce much of the nation's sweet cherry crop, fine pears, and quantities of berries. The home gardener can plant almost any temperate-zone fruit, but summers are overcast and cool, so fruits that need heat, like peaches and grapes, must be carefully chosen for the region you live in.

Zone 8b. This zone is similar to Zone 8a but with even cooler, wetter summers. It is more important in this zone to choose fruits that don't need much heat.

Climate Zones for California

Zone 7. This zone includes the high slopes above Owens Valley and the Mojave desert. The growing season of 120 to 160 days permits many fruits, but berries and dwarf trees will need heavy winter mulch to survive without damage.

Zone 8. The low desert is good for plants that need heat. Winter cold and wind make many fruit plants impossible to grow, and the summer heat rules out apples, pears, and cherries.

Zone 9. Intermediate and high desert gardeners escape the worst of the summer heat and can grow fine apples and pears and many berries. The cold temperatures rule out tender plants.

Zone 10. This zone includes the hot inland region just

This peach, surrounded by wild mustard plants, was developed for its showy flowers as well as for its fruit.

behind the Southern California coast. Commercial citrus crops are produced here. Plants that need heat do well if they can withstand some winter frost.

Zone 11. This is California's subtropical belt, with many commercial groves of avocados. Any citrus will grow in this zone. Surprisingly, the hills just behind the coast are fine for many temperate-zone fruits such as apples. Nurseries in the area carry any fruit you're likely to have heard of, and the low-chill dwarf peaches grow especially well.

Zone 12. The summer fog belt of coastal California is in many places a prime fruit area. Commercial growers around San Francisco Bay harvest European plums, Oriental plums, apricots, and sweet cherries. The coast north of the bay may have enough summer fog to affect the ripening of plants that need heat, such as peaches and grapes, so choose the hottest spots in the garden for these.

Zone 13. The interior valleys from the Inner Coast

Range to the Sierra Nevada include some of the finest fruit climates of the United States. Commercial orchardists grow peaches, plums, citrus, and strawberries, and almost all of California's wine and table grapes. High summer temperatures limit the choice of apples and pears.

Chart Symbols

A: All varieties. Plant any of the varieties that are listed in this book.

H: Hardy. Choose varieties especially developed for regions with cold temperatures.

P: Protection in winter. Use a heavy mulch over roots to protect them from freezing, or bury small fruit in mulch once the early cold has stopped all growth. Tender plants such as citrus need plastic or burlap covers over and around them in winter.

L: Low-chill. Where winters are mild you'll need spe-cial varieties of apples, peaches, and apricots. Standard varieties flower and leaf out erratically.

U: Uncertain crops. When the climate is too cold or too warm, some crops are erratic, doing well in one spot and poorly in another.

F: Fog or cool summers may affect the crop. In coastal California the summer fog cover varies. Where fog lingers most of the day, there will be insufficient heat to ripen peaches or apricots. Where fog gives way to heat in the early morning, the same crops will do well. In western Washington, rain and overcast skies have a similar effect.

(No symbol.) Fruit type is not recommended. Some crops should not be planted in the ground. There is little point in trying citrus in the mountains, for example, but you can often grow citrus in containers, sheltered in winter. If heat is the limiting factor, as in the low desert, then you'll have to do without the plants that cannot take it.

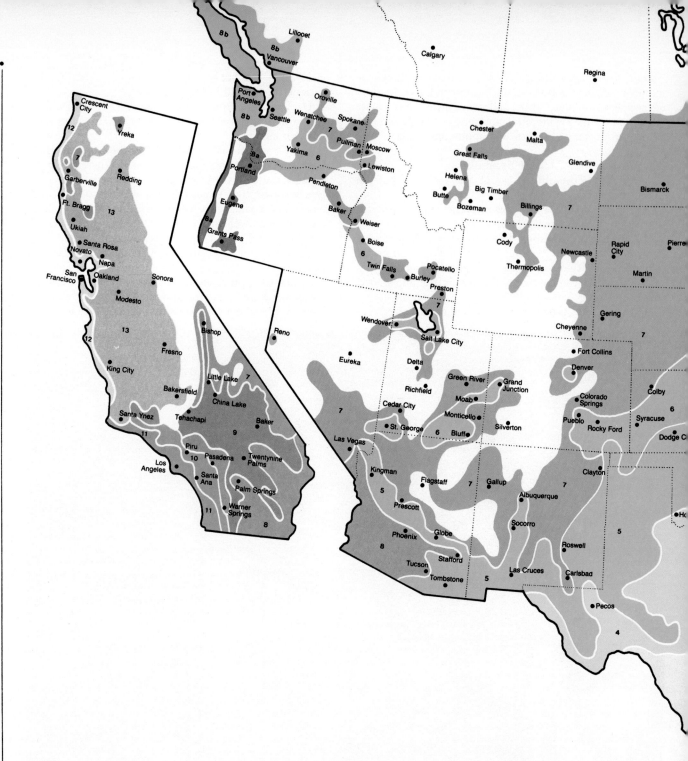

CLIMATE ZONE MAP

FRUIT	1	2	3	4	5	6	6a	6b	7	8	8a	8b	9	10	11	12	13
Apple			L,D	A,D	A	A,D	A,D	H,D	H,D		A,D	F,D	P,D	L,D	L,D	A,D	A,D
Apricot				U	H,U	H,U	H,U	H,U		L	H			L	L	A,F	A
Blackberry			A	A	A		H,P	H,P			A	A	A	A	A	A	L
Blueberry					A	A	A	P			A	A				A	
Cherry (Sweet)				U	U	P	A	P,U	P		A	A				A	A
Cherry (Sour)				U	A	A	A	H	H		A	A	A	A	A	A	A
Citrus (Hardy)	A	A	A	P						A				A	A	A,F	A
Citrus (Tender)	A	A	U,P							P			A,P	A	P,U	A,P	
Crabapple					A	A	H	H	H		U	U				U	A
Currant					A	A	A	A	A		A	A		U		A	U
Fig	A	A	A	A	P					A	H,U			A	A	A	A
Grape			A	A			A	P		U			U	A	U		A

SYMBOLS

A   All varieties
D   Dwarf varieties
F   Fog may affect fruit crop
H   Hardy varieties
L   Low-chill varieties
P   Protection needed in winter
R   Disease-resistant varieties
U   Uncertain crop
(No symbol)   Fruit not recommended

FRUIT	1	2	3	4	5	6	6a	6b	7	8	8a	8b	9	10	11	12	13
Peach and Nectarine			L,D	A,D	H,D	H	A	H	U	L,D	F	F	U,D	L,D	L,D	F	A,D
Pear			A,D	R,D	R,D	A	A	H	H		A	A	A	L,U	L,U	U	A,R
Persimmon: American				A	A	U				U	A	U	A	A	A		A
Oriental			A	A										A	A	A	A
Plum:			A	A	A	H											
European						A	A	H	H		A	A	L	U	U	A	A
Oriental						R	H			U	A	A		L	L	F	A
Pomegranate			A	A						A			A	A	A	F	A
Raspberry					A	A	A	P			A	A		U	A	A	U
Strawberry	L	L	A	A	P	P	P	P	P	U	A	A	A	A	A	A	A

*Learn the principles of fruit growth
and apply this knowledge when planting
and caring for your home orchard.
If you follow the planting, feeding,
and watering instructions in
this chapter, you'll get maximum
production in your garden.*

PLANTING AND CARE

The fruiting plants in this book have been selected because they either produce much finer crops or they tolerate certain conditions better than their wild relatives. But with all the advances brought to us by plant breeders and growers, good performance of any fruit plant still requires good care on the part of the gardener. In this chapter we discuss how fruit plants produce fruit; how they grow; and what you need to know to get them started and to keep them healthy, beautiful, and productive.

POLLINATION

You may feel that once you plant a tree you've done your part and the rest is up to nature. This idea can easily lead to fruitless plants. Before you ever lift a shovel—indeed, before you even buy a fruit tree—you need to understand a little about how fruit is produced.

Plant breeding may seem like a subject best suited to the botanist, but every time you bite into an apple or pear you're tasting the results of plant breeding, particularly the act of pollination. With a few exceptions (certain figs, for example), fruit will not form unless pollen from the male parts of a flower is transferred to the female parts of a flower. The pollinating insects for most of the fruits in this book are bees. The presence of bees around your plants, however, does not necessarily mean you'll get a crop. The pollen bees carry must be of the right sort. Most of us know that apple pollen, for example, will never pollinate a pear blossom; few of us realize that apple pollen will not always pollinate an apple blossom.

THE RIGHT COMBINATIONS

Some plants are called *self-pollinating* or *self-fertile*. This means that their flowers can be fertilized by pollen either from flowers on the same plant, or from another plant of the same kind. Self-fertile plants will produce fruit even if they are planted far away from any other plant of their kind. Among the self-fertile plants are some apples, pears, and plums; most peaches and apricots; and all sour cherries and most citrus.

Other plants set fruit *only* when they receive pollen from a plant of a *different* variety. When a plant's pollen is ineffective on its own flowers, it is called *self-sterile*. This group includes many apples; all sweet cherries; and some pears, peaches, apricots, and plums. The 'Royal Ann' sweet cherry, for example, needs another cherry tree with fertile pollen within 100 feet, or it will bear no fruit. A plant that will fertilize a sterile plant is called a *pollinator*.

Never assume that because you have a bearing fruit tree you can plant a new tree of a different variety nearby and be sure of a crop. Plants must bloom at about the same time for successful cross-pollination; for example, an early self-sterile apple will not bear fruit unless the pollinator is another early apple variety.

Some plants bear male and female flowers on separate plants. Kiwi fruit are a good example of this. You must have both a male and a female plant in your yard, and bees to pollinate the flowers, in order for the female to set fruit.

An apple orchard in full bloom. Many apples need a second variety nearby for pollination. See page 24, "Planting for Pollination".

PARTS OF A FLOWER

Bees carry pollen on body hair, then brush by the stigma to pollinate the fruit.

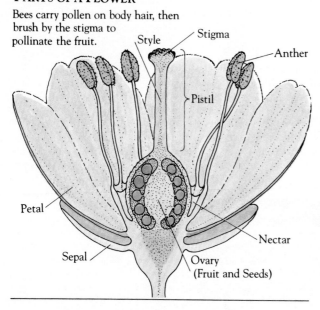

Style
Stigma
Anther
Pistil
Petal
Sepal
Nectar
Ovary (Fruit and Seeds)

PLANTING FOR POLLINATION

A fruit plant that needs a pollinator needs it close by. The maximum distance is 100 feet, but the closer the better. The bees that carry the pollen are unlikely to fly back and forth if the distance between the trees is any greater.

If your neighbor has a pollinating variety across the back fence, you're in good shape; if not, do one of the following:

☐ Plant two trees fairly close together.
☐ Graft a branch of another variety onto a tree that needs pollination.

Sometimes even though a good pollinator is near the tree that needs it and blooms at the same time, bees will fail to do the job and the crop will be poor. If this occurs, place a bouquet of flowers from the pollinating tree in a vase or jar of water, and lodge the container in the branches of the second tree. Do this early in the morning when temperatures are fairly low and the bees are hard at work.

Knowing and choosing good pollinators for your plants is a most important part of fruit gardening. The "Encyclopedia of Fruits and Berries," beginning on page 57, will tell you which varieties need pollinators and what pollinators you can use.

HOW FRUIT PLANTS GROW

All plants must have sugar to produce energy and grow. They make this sugar through photosynthesis. You can stimulate this process and provide plants with more of the sugar they need by planting them in a sunny spot, pruning and training them for good leaf exposure, keeping the soil properly watered, and keeping leaves free of dust, pests, and disease. Each piece of growing fruit needs some 30 leaves working for it, not including leaves that supply nourishment to roots and branches.

The illustration on the right gives you some idea of the day-to-day workings of a fruit plant. While the leaves are busy above ground, the roots spread out underground searching for water, oxygen, and mineral nutrients. These essential elements are then transported to the green tissues of the plant, where photosynthesis is carried out using the energy supplied by sunlight to manufacture the necessary

sugar. The sugar not immediately converted to energy for the plant's survival and growth is stored throughout the plant, including the fruit. It's easy to understand, then, that the greater the supply of factors that produce sugar—sunlight, water, and carbon dioxide—the more abundant and sweet the fruit.

SOIL

Soil should provide roots with (1) a good supply of air; (2) constant moisture (but not standing water); and (3) a good supply of mineral nutrients.

You can supply these needs best if you first examine your soil. Is it rock hard when dry and gummy when wet? If so, you have the very fine textured soil called *clay*. Clay holds moisture so well that there is little or no room for air. To correct this, aerate clay soil by adding organic matter such as peat moss or compost. Spread 4 or 5 inches of organic matter over the soil and mix it in evenly. Ideally, you should add organic material wherever the plant's roots might spread at maturity; keep in mind that the roots spread more widely than the branches.

Does water soak directly into your soil without significant spreading? Does the soil dry up a few days after watering? If so, your soil is *sandy*. Sandy soils contain a great deal of air, but moisture and nutrients wash away quickly. Additional organic matter helps here, too, by filling in spaces between the coarse soil particles and by retaining the water. Peat moss, compost, and manures are especially beneficial to sandy soils. Sawdust and ground bark are less good at holding water and actually use up nutrients as they decay, depriving the plant of them.

If you have soil that feels moist for days after watering, but still crumbles easily when you pick up a handful and squeeze it, it's just right.

DAY-TO-DAY WORKINGS OF A FRUIT TREE

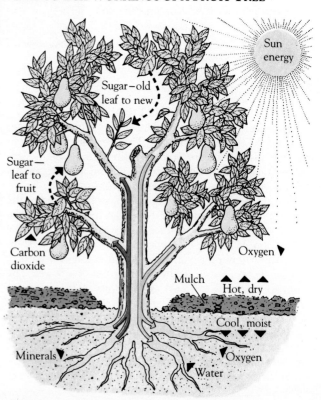

Sun energy
Sugar—old leaf to new
Sugar—leaf to fruit
Carbon dioxide
Oxygen
Mulch
Hot, dry
Cool, moist
Minerals
Oxygen
Water

Some fruits, such as pears, will tolerate dense, airless, soggy soil. Apples and crab apples will take short periods of airless soil, but apricots, cherries, figs, plums, grapes, and currants all need fair drainage. Strawberries, cane berries, peaches, and citrus need good drainage, and blueberries must have perfect drainage.

In gardens with extremely dense soil you can still plant fruits that prefer porous soils by using containers or raised beds. A raised bed for a standard fruit tree should be 3 feet deep and 6 feet long by 6 feet wide. Soils for containers are discussed on page 53.

PLANTING

Nurseries and garden centers sell bare-root plants, plants with the rootball wrapped in burlap (called "balled and burlapped"), and plants growing in containers.

Most deciduous fruit plants are sold bare root. The leafless plant is taken from the ground in late fall or winter after it has gone dormant and is shipped to the nursery, where it is held in moist sand or wood shavings. Sometimes the roots are enclosed in a plastic bag full of moist shavings. Bare-root plants are fragile and must be kept cool and moist. Plant them as soon as possible. Bare-root plants are sometimes put into containers at the nursery. If you buy them in winter or while they're still dormant, you can bare the roots again to plant them. If they have already leafed out, keep them in their containers until May or June so the root system has time to knit the container soil together.

Evergreen plants are sold either balled and burlapped or in containers made of plastic, pulp, or metal. Balled and burlapped plants are sold at the same time as bare-root plants and should go into the ground or their permanent containers quickly. Those sold in containers are available all year round and may be held until time to plant, as long as you don't cut the container.

Never let bare-root or balled and burlapped plants lie around unprotected. If you must keep bare-root plants for a time before you can plant, dig a shallow trench, lay the plants on their sides with roots in the trench, and cover the roots with moist soil. This is called "heeling in." Wrap balled and burlapped plants in a sheet of plastic so the rootball stays moist.

The illustrations here will give you some idea about how to plant a tree from the nursery. Remember never to plant if the soil is very wet. Working wet soil packs it, driving out the air and trapping the roots. In rainy climates, you can dig holes for the plants in the fall and protect the soil that's removed from drying out by covering it with a weighted plastic sheet. The soil will then be workable any time.

It's good to plant high. Notice in the illustrations that the planting soil is mounded above the normal soil line. The most fragile part of a woody plant is the crown, that section where the roots branch and the soil touches the trunk. The crown must be dry most of the time, especially in spring and fall. Raised planting minimizes crown rot (which can be fatal to the plant) by making it impossible for water to puddle near the trunk. If you plant at soil level, you invite disaster because the soil in the planting hole will settle and your plants will sink downward.

Be especially careful when planting grafted dwarf fruit trees. Dwarfing rootstocks cause dwarfing because the rootstocks are not vigorous or deep-rooted. Trees on the smallest stocks may blow over unless they have support.

PLANTING BERRIES AND VINES

Plant a rooted cutting with two or three buds above the soil. Cover these buds with mulch.

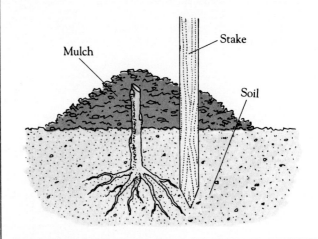

BAREROOT PLANTING

Clip off broken roots. The hole must be wide enough for the roots to spread. Soak the soil after the hole is refilled; make a volcano-shaped mound, then soak again from the top, running water slowly so it sinks in. Mound higher in dense soil, lower in good soil.

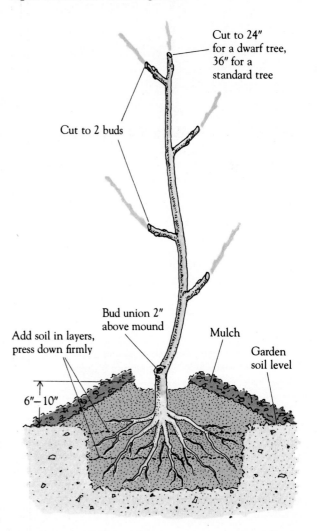

BALL AND BURLAP PLANTING

Do not jar the root ball or you may damage the tree. After the first layer of fill is pressed down, lay back the burlap and fill again. Soak, make mound, soak again. Protect all newly planted trees from winds or strong sun.

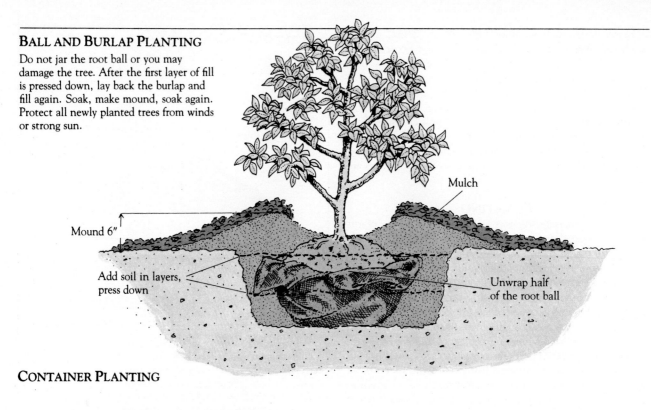

Mulch

Mound 6″

Add soil in layers, press down

Unwrap half of the root ball

CONTAINER PLANTING

Remove can before planting

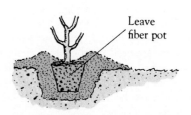

Leave fiber pot

Many growers now place the bud of the fruiting variety high on the rootstock, up to 6 or 8 inches above ground. This bud union shows later as a bulge with a healed scar on one side. Plant the tree with the union about 2 inches above the soil. This is deeper than it grew in the nursery and accomplishes two things: (1) The deep planting makes the tree a little more stable; and (2) the rootstock is less likely to send up suckers from underground. Be careful never to bury the bud union in soil or mulch at any time during the life of the tree. If moist material touches the union, the upper fruiting part will root and its vigorous root system will produce a full-sized tree instead of the dwarf you bought. Check the bud union frequently for signs of rooting and keep mulches a few inches away from it.

A good rule of thumb is to dig the planting hole twice the width of the rootball. A wise old gardener once remarked, "It's better to plant a 50-cent tree in a 5-dollar hole than a 5-dollar tree in a 50-cent hole."

FEEDING

When you feed a fruit tree, you are supplying mineral nutrients. The three primary plant nutrients are nitrogen, phosphorous, and potassium. Plants also need three secondary nutrients—calcium, magnesium, and sulfur—and very small quantities of trace nutrients, including iron, manganese, and zinc. Nitrogen in some usable form is the only element that is always in short supply. You can add it in these forms: ammonium nitrate, ammonium sulphate, calcium nitrate, complete fertilizers, or bird or animal manures.

Fruit trees rarely need extra phosphorus, but they will occasionally need potassium and the other nutrients. If growth is slow or leaves and fruit look unnatural or unhealthy, you can check with your nursery or agricultural extension agent to find out what should be added.

For a newly planted tree, use a starter solution of high phosphorus liquid fertilizer. This will encourage new root growth. Never give nitrogen fertilizer to a newly planted tree. Nitrogen will stimulate leaf growth, which is undesirable until the roots become established. Wait until the leaves just begin to expand. Keep fertilizers away from the trunk of the tree.

Although one or two feedings are often recommended, we suggest feeding equal amounts of chemical fertilizers four times at evenly spaced intervals between early spring and late June. Water very deeply after feeding. Use the following schedules for standard-sized fruit trees. Dwarf trees will require proportionately less.

First and second season. 1 tablespoon fertilizer at each feeding, scattered evenly.

Third to ninth season. Double the amount each year; for example, 2 tablespoons per feeding in the third year, 4 tablespoons in the fourth year, then 8, then 16 (one cup), and so on.

Mature tree. Continue feeding with 5 pounds ammonium sulphate; or 3½ pounds ammonium nitrate; or 6 pounds calcium nitrate; or 10 pounds of complete fertilizer containing 10 percent nitrogen.

Feeding with animal manure. Animal manure is lower in nitrogen than chemical fertilizers, and it may contain harmful salts. In dry climates be careful of bird and rabbit manures. If leaves show brown edges, soak the root area for several hours and change to another feeding method. Well-rotted cattle manure is a little safer if you water thoroughly and deeply, and it can act as both food and mulch. Some people, however, object to the odor of cattle manure.

Since manures contain less nitrogen per pound than chemical fertilizers, you can use relatively more, and because they release nitrogen slowly through bacterial action, you can put the whole amount around the tree at one time. For young growing trees, begin with a little less than ½ pound of dry bird manure or about 1 pound of dry cattle manure, and double each year. For mature trees, use 50 to 70 pounds of well-rotted bird or rabbit manure, spreading it under the outer branches in fall. For the same trees, use 100 to 200 pounds of well-rotted cattle manure.

Feeding is less a matter of exact measurement than of how the plant responds. Nitrogen forces leafy growth, but too much at the wrong time can harm your fruit crop. Feed the plant less from mid-June through leaf fall. If it produces only a few inches of growth one season, step up feeding. If it sprouts up like a geyser, feed less. And remember, dark leaves mean too much nitrogen; uniformly yellowish leaves mean too little.

WATERING

Standard fruit trees need deep watering. Dwarf trees on shallow-rooted stocks may not need as much, but they must have a constant moisture supply. At planting time, water each layer of soil in the planting hole. If the garden soil is dry, soak the hole itself before you put in the plant. Finish by soaking from the top of the planting mound, creating a depression to hold the water. Take care that water does not run over the side.

After planting, and before new growth begins, do not water again unless the soil seems dry. The roots are not growing actively at this time, and soggy soil will invite rot. When new growth begins, let the top inch or so of soil dry; then give the plant a thorough soaking. Be sure to water at the top of the planting mound. This is especially important with balled and burlapped plants, since the soil in the rootball may not take up water unless it is applied directly overhead.

When first-season growth is abundant and plants are growing well in midsummer, stop watering from the top of the mound. Dig a shallow ditch around the base and soak the soil about every 1 to 3 weeks, or whenever the top inch or two of soil dries out.

After the first season make a shallow ditch about 6 to 12 inches wide around the plant and just outside the tips of the branches. Move the ditch outward as the plant grows. Soak thoroughly about once every 3 or 4 weeks. This is a rough guide. Your tree may need water more often in very sandy soil, less often in heavier soil. Always dig down a few inches into the soil first to see if watering is necessary.

Your aim in watering is to soak the soil long enough for moisture to reach the deep roots. Remember that water sinks quickly through sandy soil, but very slowly through clay. On standard trees the deepest roots may penetrate the soil to a depth of many feet. On dwarf trees the deepest roots may extend only 30 to 36 inches beneath the surface. You can check your watering by pushing a stiff wire 4 to 5 feet long into the wet soil. It will penetrate only wet soil, so when you cannot push it down any farther, pull it out and check the depth of penetration. Soil should be moist down to at least 2½ feet for dwarfs, 3½ to 4 feet for standard trees.

Trees in a lawn area should have a deep soaking about twice a summer in addition to normal lawn watering.

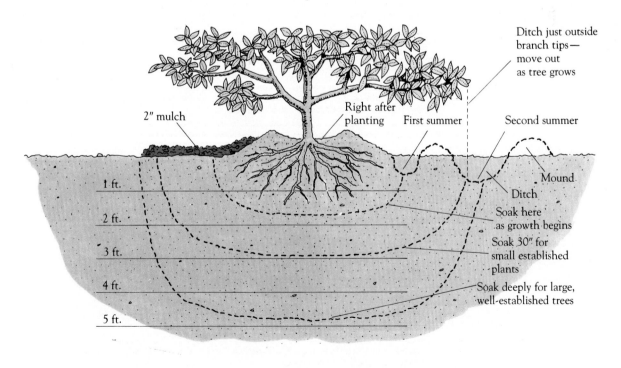

2″ mulch

Right after planting First summer Second summer

Ditch just outside branch tips— move out as tree grows

Mound
Ditch
Soak here as growth begins
Soak 30″ for small established plants
Soak deeply for large, well-established trees

1 ft.
2 ft.
3 ft.
4 ft.
5 ft.

MULCHING

Mulch is simply a cover over the soil. It may be gravel, sheet plastic, ground fir bark, or any of a number of organic materials. Organic mulches are particularly valuable, because as they break down they improve soil texture and add nutrients. The major advantages of mulching are that it:

- ☐ Keeps wind and sun from baking and crusting soil
- ☐ Smothers weed seeds
- ☐ Protects the soil from pedestrian traffic
- ☐ Holds in water even at the surface
- ☐ Keeps surface soil cool in summer (organic mulch), warm in winter (plastic mulch)
- ☐ Keeps soil from alternately freezing and thawing, which can heave plants from the ground

The chart below lists many mulching materials, along with some of their advantages and disadvantages.

Applying Mulch

In mulching, cover the soil with at least 2 inches of a porous material extending from the planting mound out to a line directly under the branch tips. If you use a waterproof mulch like thick paper or plastic, slash it at intervals so that water can pass through. You can cover plastic mulch with gravel, pebbles, or bark chips spread over the surface about an inch deep. Hide paper mulches with a thin layer of ground bark.

In warm, dry climates, spread mulch in spring and then dig it into the top 2 inches of soil in late fall. In cold climates where the ground freezes, add more mulch in late fall, up to 6 inches deep. You may even want to cover this deep mulch with evergreen boughs for added protection and to anchor the mulch in strong winds. Snow increases the protection afforded by any mulch.

Never pile mulch against the tree trunk or plant stem; keep it at least 6 inches away. Wet mulch can cause rot, and

PROPER MULCHING METHOD

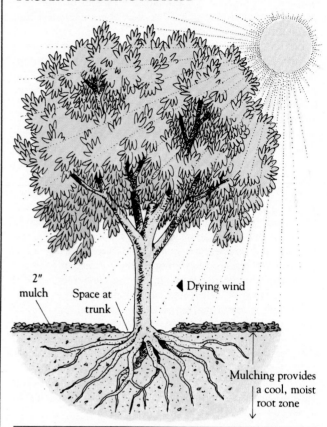

2" mulch Space at trunk ◀ Drying wind

Mulching provides a cool, moist root zone

dwarf trees mulched to the bud union will take root and will grow to full size. The one exception to the rule is in very cold climates, where it is good to mulch deeply over the bud union when real cold begins. However, be sure to remove the mulch when severe cold is over.

MULCHING MATERIALS

Material	Remarks
Rotted manure	May contain weed seeds.
Sawdust Wood chips Wood shavings	Low in plant nutrients, decomposes slowly, tends to pack down. Well-rotted material preferred. Can be fresh if nitrate of ammonia or nitrate of soda is added at the rate of 1 pound per 100 sq. ft. Keep away from building foundations; may cause termites.
Peat moss	Attractive, available, but expensive for large areas. Should be kept moist at all times.
Ground corn cobs	Excellent for improving soil structure.
Pine needles	Will not mat down. Fairly durable.
Peanut hulls Cotton screenings Tobacco stems (shredded)	Supply plant nutrients and improve soil structure. Fairly durable.
Tree leaves (whole or shredded)	Excellent source of humus. Rot rapidly, high in nutrients.
Hay or grass clippings	Unattractive, but repeated use builds up reserve of available nutrients which lasts for years.
Straw	Same as above, but lower in nutrients although furnishes considerable potassium.
Buckwheat hulls	Very attractive but tend to scatter in windy locations.
Pecan hulls	Extremely durable, availability limited.
Gravel or stone chips	Limited use, but particularly good for rock garden plantings. Extremely durable, holds down weeds, but does not supply plant nutrients or humus.
Bark	Ground and packaged commercially. Especially attractive in this form. Sometimes available in rough form from pulpwood loading sites.

PESTS AND DISEASES OF FRUIT

The more energy your plants expend recovering from the effects of pests and diseases, the less fruit they will bear. Here are some tips on giving them a helping hand that will bring ample reward at harvest time. Whenever using a pesticide be sure to spray as directed by the product label.

The pests listed here are among the most common, although you may encounter others. A few are confined to specific regions of the country.

Apple Maggot (Railroad Worm)

Primarily found east of the Rockies, the apple maggot has adult flies that lay eggs under the skin of the fruit, and larvae that hatch and tunnel through the flesh. The flies are active from July through harvest. Keep trees clean and remove damaged fruit. Spray with diazinon or carbaryl products, following label directions. Do not use diazinon within 14 days of harvest.

Birds

Birds are at their worst with cherries, blueberries, and other fruit that they can remove entirely. When fruit begins to ripen, cover the entire plant with plastic netting, available from nurseries and hardware dealers. Throw the net directly over the plant, or build simple wood frames to support netting over dwarfs and bushes. For larger trees, cotton twine may be sufficient discouragement. Throw a ball of it over the tree repeatedly from different sides. The strands annoy the birds when they try to land. The twine usually rots away over winter.

Cherry Fruit Fly

Found in the Pacific Northwest, the cherry fruit fly starts as a white larva that burrows through the cherries, leaving a hole. As soon as you notice flies or maggot damage, spray at 7-day intervals with diazinon or any product registered for control of this pest. Do not spray diazinon on the tree within 14 days of harvest.

Cherry (Pear) Slug

These small, wet-looking green worms are the larvae of a wasp. They skeletonize leaves, leaving lacy patches. Spray when you notice them (probably in June and again in August) with malathion or a contact spray registered for control of the pest. Follow label directions. One spraying is usually enough.

Codling Moth

The major apple and pear pest, these moths lay eggs in the blossoms, and their larvae tunnel in the fruit, leaving holes and droppings (frass). Spray with diazinon or an insecticide containing malathion and methoxychlor after petals fall, and continue to spray every 2 to 4 weeks as directed.

Flatheaded Borer

This western pest is the larva of a beetle. The borer burrows into bark that has been damaged or sunburned. To avoid the pest, avoid the damage: Paint or wrap trunks, and be careful not to cut them with tools or machinery. When you find tunnels and droppings, cut away bark and wood and dig out the borers. Be sure you have eliminated them all, and then paint the wound with tree seal or asphalt emulsion. For apples, cherries, and pears, use a lindane product registered for control of this pest.

Leaf Roller

This moth larva hides in rolled leaves and feeds on both the foliage and fruit. Once established, it is protected from spray. Infested leaves must be picked off. Spray with diazinon, carbaryl, or malathion when pests first appear and continue spray treatments according to the intervals recommended on the label.

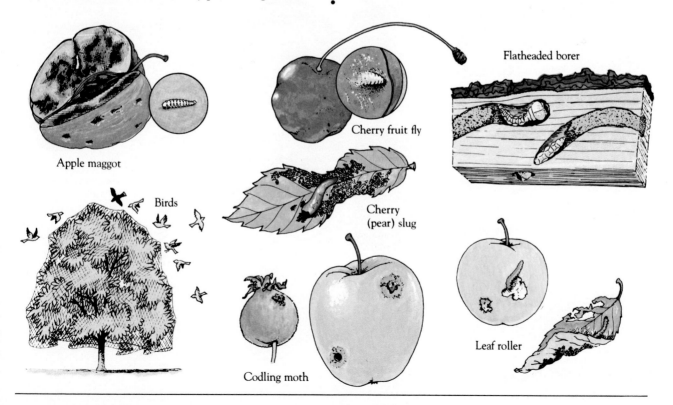

Apple maggot

Birds

Cherry fruit fly

Cherry (pear) slug

Flatheaded borer

Codling moth

Leaf roller

Mites

Spraying for insects may trigger a mite attack, because spraying kills the mites' enemies. You'll know mites are present by a silvery webbing under the leaves, or if the leaves are curled, stippled, or bronzed. Kill overwintering mites with dormant oil spray. During the growing season, use kelthane or other products registered for mite control on fruit trees.

Peach Tree Borer

There are two kinds of peach borer: One bores into twigs (peach twig borer); the other very common borer attacks the trunk at the soil line. Dig soil away from the trunk and check for tunnels and droppings. Kill the worm by pushing a bit of wire down its tunnel. Spray trees with diazinon, carbaryl, or lindane products according to the directions on the label. Check cherry and plum trees for infestations of the same pest.

Pear Psylla

The larvae of this pest cluster on leaves like aphids and suck plant juices. They excrete a sticky, sweet honeydew that coats the leaves and fruits. A black, sooty fungus may grow on the honeydew, reducing photosynthesis and weakening the tree. Dormant oil spray is a very effective treatment against pear psylla. Early stages of infestation can be controlled with diazinon sprays, or sprays containing malathion and methoxychlor.

Plum Curculio

A serious problem for many stone fruits east of the Rockies, this pest belongs to the beetle family. Both the adults and larvae damage the fruit. Spray with an insecticide containing malathion and methoxychlor or other products labeled for control of this pest.

Rodents

Mice, voles, and rabbits all like the bark of young trees, especially when it is covered with mulch or snow in winter and better food is unavailable. If enough bark is removed, the tree will die at the first growth surge of spring. Protect the lower trunk in winter or year round with a cylinder of hardware cloth. Be sure the cloth doesn't become tight as the tree grows. Check during the season and loosen or replace it.

Rosy Apple Aphid (and other aphids)

Aphids are soft-bodied insects that damage leaves and fruit by sucking plant sap. A dormant oil spray kills overwintering eggs, and a contact spray of malathion or diazinon helps control the insects during the growing season.

San Jose Scale

This pest causes red spots to develop on infested fruit. San Jose scales appear in masses when infestations are heavy. They can kill a plant in one or two seasons. Use a delayed dormant oil spray to control mature scales before crawlers hatch or are born. Crawlers can be controlled with malathion or diazinon sprays applied in May or June.

Tent Caterpillar

You probably won't see this pest if you have sprayed early for others. Tent caterpillars build large webs among branches; these webs contain hundreds of hairy caterpillars that emerge to eat leaves. Use a product registered for control of tent caterpillars on fruit trees.

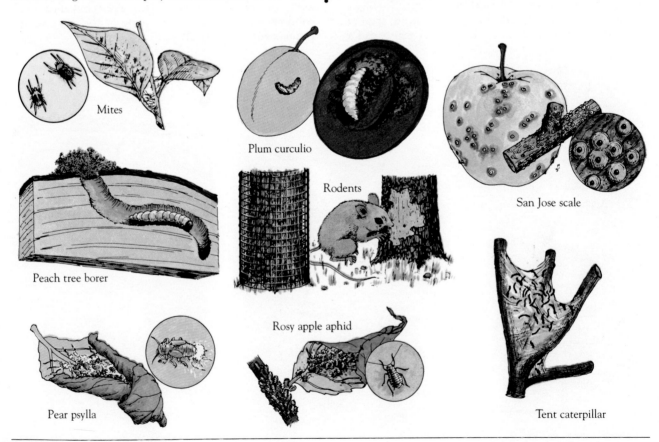

Mites

Plum curculio

Peach tree borer

Rodents

San Jose scale

Rosy apple aphid

Pear psylla

Tent caterpillar

DISEASES OF FRUIT TREES

The following are the most common of the many fruit tree diseases. Some are easily controlled with proper sprays (the right season is very important); others are best fought by choosing resistant plants; and some require removal and burning of infected parts.

Any disease is easier to deal with if trees are kept pruned and the ground around them is cleared of fallen fruit and leaves. Pick shriveled fruit that hangs on and either burn it or seal it in a bag to be discarded.

Be sure to prepare soil properly, plant high, and keep the tree watered and fed. A healthy plant is much more resistant to disease.

Apple Scab

This disease overwinters on leaf debris, so adequate garden cleanup is important. The fungus infects foliage and fruit. The disease is most severe in wet weather. Scab is not a problem in dry-summer areas. Both apples and crab apples are affected and need spraying. Choose resistant plants when possible.

To control scab, apply captan, dodine, lime sulfur, or benomyl to apples at regular intervals, before and after the tree blooms.

Bacterial Leaf Spot

Primarily attacking cherries, peaches, and plums, this bacterial disease lives through winter in leaf debris. Infection occurs during rainy periods in the spring, causing spots to form that turn brown and develop into widening holes in the leaves. It may attack fruit spurs and cause fruit drop.

This disease cannot be eliminated, but it may be suppressed by spraying with basic copper sulfate at petal fall.

Bacterial Canker of Cherries

The disease causes long, narrow, damp-looking gum-edged patches on trunk or branches. Branches die as they are girdled. In wet climates avoid 'Bing', 'Lambert', 'Napoleon', and 'Van'. Resistant varieties are 'Corum' and 'Sam'. The disease can also affect apricots, blueberries, peaches, and prune plums. Do not use susceptible peaches in damp climates.

Brown Rot

This disease is serious on all stone fruits, but especially on apricots, peaches, and nectarines, in some regions making them nearly impossible to grow to the edible stage. The disease causes blossoms to brown, turn wet-looking, and drop. Brown rot eventually causes fruit rot on the tree. Infected fruit must be removed by hand to prevent reinfection.

To control the blossom-blight phase, spray as the first pink shows, using a fungicide such as captan, benomyl, dodine, ferbam, ziram, or lime sulfur.

To control attacks on fruit, spray as the fruit begins to ripen (green fruit is rarely attacked) and repeat if there is a period of wet weather. On peaches and nectarines, the disease may attack twigs and overwinter on them. Pruning out dead twigs controls the disease in the following year.

Cedar Apple Rust

This disease appears only where the alternate host, certain species of juniper or red cedar, grows near apples. The leaves first show orange spots and odd, cup-shaped structures; then they turn yellow and fall. Remove junipers and red cedars, or avoid planting them. If you have ornamental cedars, remove the galls in summer. Galls are brownish and globe-shaped, and look like part of the tree. Spray the apples with

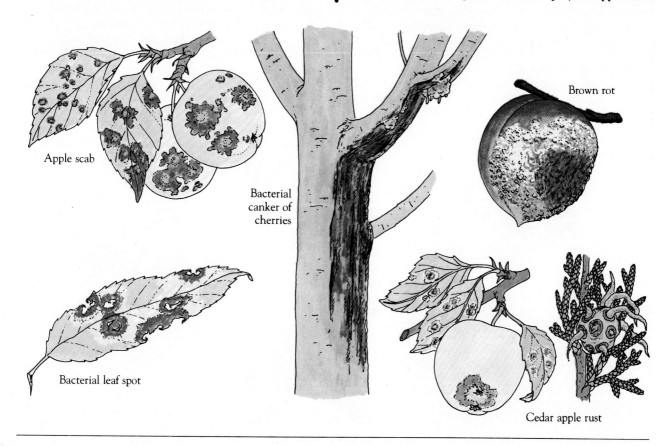

Apple scab

Bacterial canker of cherries

Bacterial leaf spot

Brown rot

Cedar apple rust

a product registered for control of cedar apple rust (ferbam or zineb products), following directions carefully.

Crown Gall

This disease of bacterial origin occurs in many soils. It attacks young trees, producing soft, corky galls or swellings on the crown and roots. The galls often grow until they girdle and stunt the tree.

Avoid buying young trees that show galls, and plant young trees carefully to avoid injury, as injury allows bacteria to enter the plant. Older trees with galls can be treated with a product called Gallex. Recent experimental work has indicated that young plants can be successfully inoculated against the disease.

Crown Rot

This is a common serious disease of almost any plant growing in a soil that is frequently or constantly wet. Crown rot is usually caused by a fungus. Infected branches redden and foliage yellows or discolors. Look at the bark below the soil line to see whether it is dead, and if so, scrape the bark away and pull back the soil so that air can reach the infection.

Avoid crown rot on established trees by planting high and watering well out from the trunk. The soil should never be wet at the crown.

Fireblight

Fireblight is spread by insects during the bloom period and shows later in spring as new growth wilts, turns dark, and finally blackens as if burned. The infection spreads quickly down branches, eventually killing infected pears and damaging or killing apples.

No treatment has proven completely effective for the home gardener, other than severe pruning and burning of infected wood. Choose resistant pears. Where the disease is severe, plant resistant apple cultivars such as 'Cortland' or 'Delicious'. Mild winters will increase the severity of the disease the following spring.

To control infection on resistant plants, cut off any blighted branches several inches below the infection as soon as you notice an attack. There is some question about the best method of sterilizing pruning tools, but an effective method is to carry a container of rubbing alcohol with you and to dip the shears after each cut. Burn or remove all prunings immediately.

Gummosis

Deposits of gum on stone fruit branches is fairly common and may occur because of mechanical damage, insect damage, or as a result of a number of diseases. Several serious bacterial diseases with this symptom almost rule out the planting of certain fruit varieties in some areas. There is no spray treatment. A disease that often causes gummosis is bacterial canker of cherries, described on page 31.

Peach Leaf Curl

This disease attacks peaches in many regions, interfering with blossom and fruit production. The disease shows up as a reddening of leaves; they then curl into blisters, which may have a powdery look, and finally they fall. A second crop of leaves grows, which is not affected. No spraying is effective once the disease has appeared, but a basic copper sulfate spray or lime sulfur will control it completely if you spray every year. Uncontrolled, it will weaken trees and will kill them after a number of seasons.

To control curl, you should spray immediately after leaf drop, or just before the buds break, with a fixed copper spray or lime sulfur, wetting every twig and branch completely. If rain falls immediately after treatment, repeat the treatment.

Powdery Mildew

Powdery mildew is a fungus that causes a grayish, powdery coating to form over young shoots, leaves, and flower buds, often deforming or killing them. It thrives in shade and where air circulation is poor. Be sure that fruit plants receive sunlight most of the day and grow where air moves freely.

Where an infection begins, clip off severely mildewed twigs and spray the tree with cyclohexamide or dicofol.

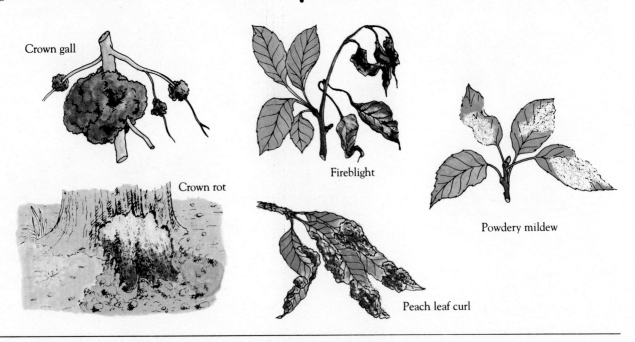

Crown gall

Crown rot

Fireblight

Powdery mildew

Peach leaf curl

BASIC FRUIT PEST CONTROL

Complete control of insects and diseases on fruit crops requires a thorough and comprehensive spray program. Proper timing, good coverage of foliage, and correct chemicals are essential.

Here is a simplified spray program that should meet your essential needs. But keep in mind that some pests are difficult to control, so you may find some blemished fruit even after following these recommendations. Your neighbors' spray programs are important, too; if they are negligent, then you may find your neighbors' pest problems becoming your own.

Apples and Pears

Winter: Before leaves are out, apply a dormant oil spray to control scale, mites, and other pests. This can be the most important spray of the year. Follow directions fully and completely for best results.

Spring: The next important spray is when fruit buds show pink at the tips. Spray with an insecticide such as diazinon to control aphids, leaf rollers, and many other pests and with a fungicide such as captan to control apple scab, fruit spot, and bitter rot. When three-quarters of the petals have fallen, spray an insecticide again to stop codling moth, which appears at this time. Where apple scab is a problem, consistent use of a fungicide such as captan is necessary for control.

Summer: The first summer spray is 10 to 14 days after petal fall. Use an insecticide, fungicide, or both, as your trees require. For perfect fruit, continue to spray through the summer with insecticide and fungicide as needed.

Fall: Spray as necessary, but pay strict attention to label instructions regarding time intervals between spray treatments and harvest.

Peaches, Apricots, Cherries, and Plums

Winter: Same treatment as with apples and pears above. Where peach leaf curl is a problem, timing is very important. Lime sulfur sprays such as 26 percent calcium polysulfides should be applied in October or November after leaf drop. In areas of heavy winter rainfall, do this before rains begin. Apply another full-coverage spray in early spring before buds begin to swell. Note: If buds have begun to swell or open, it is too late to obtain satisfactory control, as infection has already occurred. For best results, spray in fall and early spring.

Spring: When blossom buds show color—pink for peaches, red for apricots, popcorn stage for cherries, and green-tip stage for plums—spray an insecticide such as diazinon to control insect pests. A fungicide such as captan is often needed for brown rot control. Spray an insecticide and fungicide again when three-quarters of the flower petals have fallen.

Summer: Same as above.

Fall: Same as above.

Direct the spray to all sides of the leaves. Aphids are usually on new leaves, mites live on the bottoms of leaves, and scale insects are usually on stems and branches.

*Pruning encourages healthy growth
and larger fruit, while
training conserves space and makes
harvesting easier.
With just a bit of practice at both
techniques, you'll feel like a pro.*

Pruning and training are two operations of extreme importance to the gardener who grows fruits. There is no clear division between these two operations. Most training involves a lot of careful pruning and a little bit of actual training. We give you instructions for correctly pruning fruit plants, and then show you how these techniques apply to training plants into functional shapes. Once you have mastered the basics, you can experiment with techniques for training into more formal shapes.

PRUNING

Is it really necessary for the home gardener to learn how to prune? Most of us have seen long-neglected apple or pear trees or unpruned tangles of blackberry vines that still bear delicious fruit. This suggests that pruning isn't really essential or that we can get by without it.

Plants will live, grow, and bear fruit without ever being pruned, but experience has shown that good pruning can prevent or remedy many of the problems that arise in some plants. Pruning is probably best viewed as the most effective means to head off trouble, improve your plants' performance, and keep them in excellent condition.

Pruning means simply removing part of a plant to benefit the whole. When you cut away any part of a plant, it directly affects the plant's growth. Depending on how and when it is done, pruning can:

This espaliered apple takes up almost no room in the garden. See page 47 for information on dwarfing and training apples.

☐ Produce new growth where it is desired
☐ Help control growth
☐ Shape a young plant
☐ Correct or repair damage
☐ Help control insects and diseases
☐ Rejuvenate or reshape an older plant
☐ Bring about earlier blooming
☐ Increase the production, size, and quality of fruit

These advantages make pruning well worth undertaking, even if you are inexperienced and, like many, timid about cutting your plants.

Good pruning requires knowledge, foresight, and care. As a rule of thumb, never make a cut without a clear idea of its probable effect on the plant. At the same time, however, don't be so fearful of cutting that you can't get the job done. If you keep in mind that proper pruning is beneficial to plants, and then proceed carefully, you'll get good results.

Getting Started

No job can be done well without the right tools. To start, you'll need a good pair of pruning shears, and if you plan to be making cuts larger than shears can handle, you'll need more tools. See page 36. It's worth noting that the best-quality pruning tools are generally more expensive, but are worth the extra cost in the long run, because they will last much longer than inexpensive tools.

If you find yourself in a quandary about where to begin, remember that you can't hurt a plant by cutting out dead, diseased, or damaged wood or wood that crosses and rubs against other wood (which can cause wounds susceptible to infection). On the contrary, you'll be doing a great deal of good. Eliminating these problems is the place to start for inexperienced and experienced pruners alike. It will provide a clearer view of the tree and of the remaining work to be done, and will open up the tree to more air and light.

Plants vary in their needs for pruning. Some need a good deal of pruning every year; some need only a little in an entire lifetime; and some never need pruning unless injured. No other plants in the garden depend on pruning as much as fruit trees. Unfortunately, no trees vary as widely in the most effective means of pruning. If you want to grow beautiful apples or peaches, you must learn the difference between pruning an apple and a peach tree.

The objective in pruning a fruit tree is to produce an abundance of good-quality fruit throughout the branches, including the lower and interior ones. An unpruned peach or nectarine tree will bear mostly on branch tips, and the leaves that produce the sugars that nourish the plant and accumulate in the fruit will grow only where they receive sufficient light. An unpruned tree with heavy growth in its interior, restricting light and air circulation, is more susceptible to fungus diseases. In addition, heavy loads of fruit at branch tips can make harvest difficult and cause splitting and breaking branches, wounds that invite pests and disease. To head off these problems and produce a more even distribution of leaves and fruit, you need to thin out the tree by removing selected branches. The increased light and air circulation achieved by this pruning will help prevent fungus diseases and the open tree will be much more accessible to thorough coverage with any necessary sprays. The tree will also bear more uniformly and will be easier to harvest. Clearly, then, one of the main purposes in fruit tree pruning is to cut to admit light and air to interior leaves and fruit.

The illustrations on the next page will help you recognize the various parts of a plant that you need to know when you get to work.

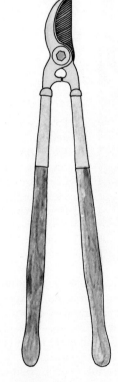

Hand pruners:
Use on stems up to ¾″ in diameter. There are two main types:

☐ Scissor-style pruners have sharpened blades that overlap in making the cut.
☐ Anvil-style pruners have a sharpened top blade that snaps onto a flat plate of softer metal.

Lopping shears:
Use on branches up to 1¼″ in diameter. Heavy duty loppers are available for cutting through wood 1¾″ thick.

Narrow curved pruning saw:
Use on branches up to 2″ in diameter where the branches are too densely crowded to effectively wield a wide blade saw.

Wood rasp:
Use to smooth the rough edges of very large cuts on trunks or thick branches.

Extension saw:
Use on out-of-reach branches up to 2″ in diameter. The curved blade operates like the narrow curved pruning saw.

Pruning paint:
This spray-on seal is used on pruning cuts over ½″ in diameter.

Pruning compound:
Use the built-in applicator to apply the seal on any cuts over ½″ in diameter.

Pruning knife:
Use to smooth the rough edges of a cut on a trunk or a branch. Smoothing the edges helps the tree heal quickly.

ANATOMY OF A FRUIT TREE

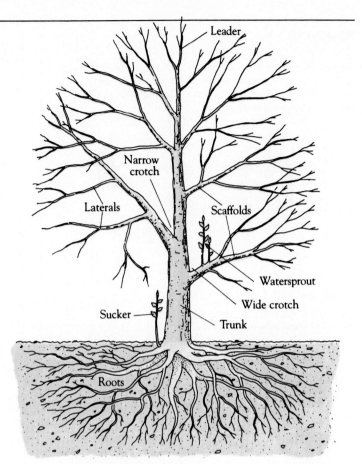

Crotch: A crotch is the angle where branches fork, or where a main limb joins the trunk. Strong crotches are wide angled—45 degrees or more: Weak crotches are narrow.

Scaffold: The main limbs branching from the trunk.

Watersprout: A very vigorous shoot from a dormant bud on an old branch. Remove by cutting at the base.

Sucker: A vigorous shoot from the roots or from below the bud union. Cut off at the base.

CHOOSING THE RIGHT BUD:

Prune to the lateral bud that will produce the branch you want. An outside bud will usually produce an outside branch. The placement of that bud on the stem points the direction of the new branch.

PARTS OF THE BRANCH

Terminal Bud: The fat bud at a branch tip will always grow first and fastest if you leave it. Cut it, and several buds will grow behind it.

Leaf Bud: Flattish triangle on the side of a branch. To make one grow, cut just above it. Choose buds pointing outward from the trunk so the growing branch will have space and light.

Flower Bud: Plump compared to leaf buds and first to swell in spring. On stone fruits they grow alone or beside leaf buds. On apples and pears they grow *with* a few leaves.

Spurs: Twiglets on apples, pears, plums, and apricots. They grow on older branches, produce fat flower buds, then fruit. Don't remove them.

Bud Scar: A ring on a branch that marks the point where the terminal bud began growing after the dormant season. The line marks the origin of this year's growth.

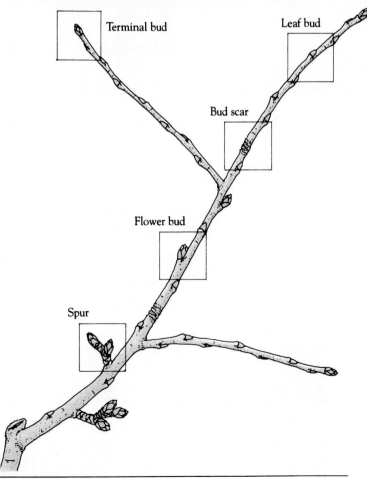

REMOVING A STUB:

Never leave a projecting stub. It will rot and can damage the branch it is attached to. Cut stubs close to the trunk at a point where the wound will be about the same diameter as the branch you cut. Cutting very close leaves a larger wound.

ANGLING THE CUT:

Cut at an angle about ¼" above a bud or leaf. As the bud grows, new bark will cover the raw wood.

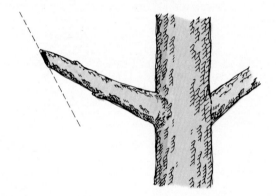

CUTTING A LARGE LIMB:

When you remove a big branch, first undercut at a short distance from the trunk (A). Then saw off the branch beyond the undercut (B). Finally, cut the stub close to the trunk (C). This technique will prevent a falling branch from tearing the bark. Paint the wound with a pruning compound.

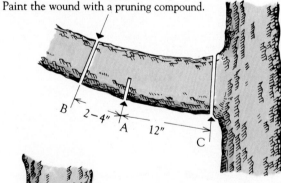

A WELL-HEALED CUT:

New tissue grows from the edge of a wound toward the center. If your cut was smooth, the scar will close evenly with a dimple in the very center.

Making a Cut

When you cut away part of a plant, you leave a wound susceptible to pests and diseases. To avoid trouble, always try to make wounds as small as possible. If they are more than ½ inch in diameter, it's a good idea to protect them with pruning paint or commercial pruning sealer.

The smallest possible wound is made by removing a bud or twig. If a new sprout is growing in toward the center of the tree or toward the trunk, or threatening to tangle with another branch when it grows longer, pinch it off now, to save pruning later. If you see the bud of a sucker down near the soil, rub it off with your thumb.

Always make cuts close to a node. Branches grow only at these nodes, and if you leave too long a stub beyond the node it will die and rot. Be sure to cut at a slight angle so that there is no straight "shoulder" left to attract disease or burrowing pests.

PRUNING METHODS

During the first three years of cultivation, most fruit trees can be pruned in the same way. When they begin to bear, however, each species should be pruned differently. All bearing fruit trees can be pruned annually, with additional light pruning in the summer to expose fruit spurs.

Commercial fruit growers prune fruit trees in three ways, and each has its own advantages. Some growers use the older method called *vase pruning*. Another popular method is *modified central-leader pruning*. The third method, *delayed open-center pruning*, combines both techniques. Remember that dwarf trees will require less severe pruning because they are smaller.

Vase Pruning

In this method the tree is shaped to a short trunk of about 3 feet with three or four main limbs, each of which has fully filled-out secondary branches. This shape creates an open center allowing light to reach all branches.

Vase pruning is always used with apricots, plums, and peaches and often with pears and apples.

Modified Central-leader Pruning

In this technique the tree is shaped to one tall trunk that extends upward through the tree, clearly emerging at the top. This shape makes a strong tree, but since the center is shaded, less fruit is produced. The smallest dwarf apples are pruned in this shape in a variation called the "spindle bush." Because the tree is as small as a bush, shade and pruning are not problems.

Delayed Open-center Pruning

This method produces both the strength of a central trunk and the sun-filled center of a vase-shaped tree. A single trunk is allowed to grow vertically until it reaches 6 to 10 feet tall. It is then cut off just above a branch. Main scaffold branches are then selected and pruned to form a vase shape. Subsequent prunings follow the vase method.

Productive fruit trees require that a definite method of pruning be established from the moment the tree is planted. Such pruning and training will keep the tree balanced in form and—very important—balanced in new and young wood. Left unpruned, the tree will become dense with weak, twiggy growth and overloaded with small, less healthy fruit.

THREE METHODS OF PRUNING

Vase pruning

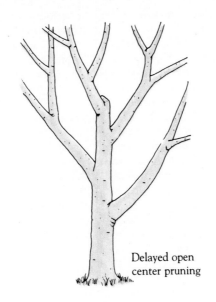

Delayed open
center pruning

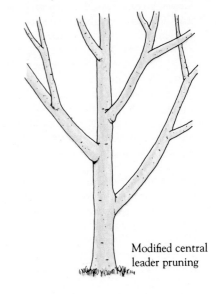

Modified central
leader pruning

HOW TO PRUNE FOR A VASE SHAPE

The general requirements for establishing a vase shape are given here, but be sure to consult the specific procedures in the following section for each type of fruit tree.

When you plant a young bare-root tree, it will normally consist of a thin vertical shoot, called a *whip*, and some twiggy side branches. To start the tree on a good course, cut the whip back to about 2 feet above the ground for a dwarf, 2½ feet for a standard tree. Cut just above a bud and then prune any side branches back to 2 buds.

First Dormant Season

After the new tree has grown through the first spring, summer, and fall, into its first winter dormancy, choose three or four branches with wide crotches, as shown in the illustration. Examining the tree from above, look for branches that radiate evenly around the trunk with almost equal distance between them. You should also try to have at least 6 inches vertical distance between branches, with the lowest branch about 18 inches above the ground. If there are three such branches, cut off the vertical stem just above the top one. If there are fewer than three good branches, leave the vertical stem and choose the remaining scaffold branches during the next dormant season.

Second Dormant Season

If still necessary, choose remaining scaffold branches and cut off the vertical stem just above the highest selected scaffold branch. The scaffold branches you chose during the first dormant season will have grown side branches. Remove the weakest of these, leaving the main stem and laterals on each branch. *Do not prune twiggy growth.*

Third Dormant Season

Now is the time to thin surplus shoots and branches. Select the strongest and best-placed terminal shoot near the tip of each scaffold branch, as well as one or two other side shoots on each branch. Remove all other shoots on the branch. Leave the short weak shoots that grow straight from the trunk to shade it and help to produce food for the tree.

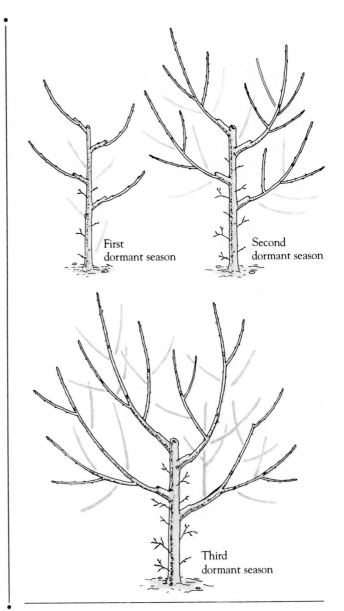

First
dormant season

Second
dormant season

Third
dormant season

Mature Trees

After the third season in the ground, a fruit tree will need only light thinning until it begins to bear.

About a month before the fruit is to be picked, *summer pruning* can be done to maintain the correct shape of the tree (see page 48). Shorten new shoots that threaten to destroy the shape and that block out too much sun.

It is always necessary to *thin the fruit* if you want large, top-quality fruit. Otherwise nature will produce the most fruit in order to get the largest number of seeds to perpetuate the species. The tree has only so much food and energy, and you have a choice of one large fruit or two small ones. To see how much difference thinning makes, leave a branch unthinned and compare its fruit at harvest time to that of a properly pruned branch.

Although each variety has a best time for thinning, a good rule of thumb is to thin before the fruit has gone through half its growing season. Apricots should be thinned to about 2 to 4 inches between fruit; and peaches, nectarines, and the larger plums to about 4 or 5 inches. For the biggest apples, leave about 6 to 8 inches of space and 1 apple per fruit spur. Cherries are rarely, if ever, thinned.

SPECIFIC PRUNING PROCEDURES

The following are pruning methods for each kind of fruit tree and berry bush listed in the "Encyclopedia of Fruits and Berries" beginning on page 57. Additional special tips on pruning fruits and berries appear in the individual entries of the encyclopedia.

Apple Trees

Vase pruning has been the method commonly used by orchardists to train standard apple trees. After you have picked your scaffold branches, as described earlier, cut them back one-third to encourage a strong branch system near the trunk. In the second and third dormant season, reduce the length of all new growth by one-third, and thin out to create a strong, evenly spaced framework of branches. These secondary scaffold branches are the ones that will develop fruit spurs on their lateral branches. The pruning during this period should always be to a bud on the top of a branch that points outward. This will develop the vase shape.

The modified central-leader system is not recommended for a standard apple tree.

SPINDLE-BUSH PRUNING

For dwarf apples on Malling rootstocks

When you plant: Cut the whip 20″ above the ground. Trim branches to two buds. Place a 6′ stake on the windward side and tie the tree loosely.

First dormant season: Choose 3 or 4 branches evenly spaced around the trunk, and all growing at about the same height above the ground. If a crotch is narrow, tie a weight to the branch to pull it outward.

Remove any side branches on the next 18″ of trunk. Do not cut the vertical stem.

Second dormant season (not pictured): Choose a second circle of branches growing close together about 20 to 24″ above the first circle. Leave three or four, evenly spaced around the trunk. Cut side branches above these, but not the vertical stem. Trim off the weakest side shoots on the lower circle, leaving the best ones and any twiggy growth. Retie the tree loosely.

Third dormant season: Cut all growth along central stem for about 20″ above the second circle of branches. In succeeding years, prune to maintain a triple tier, but remove as little growth as possible.

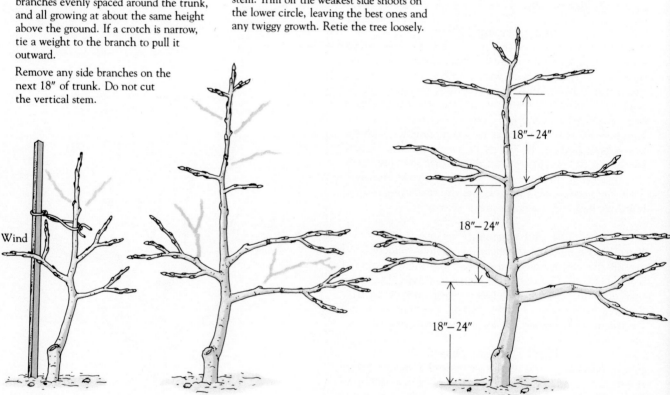

With semidwarf and dwarf apple trees, delayed open-center or vase pruning can be used, but the modified central-leader system makes these trees stronger and earlier bearing. When planting the bare-root dwarf, cut back all branches, including the top, about one-quarter, or about 8 to 10 inches. Make cuts to a strong outside bud. For the second and third year, repeat the process to train the central leader up and the scaffold branches out, parallel to the ground. Most dwarfs will begin to bear the second and third years and will bear heavily thereafter.

For dwarf apples on Malling 9 rootstock, spindle-bush training is effective. See illustration on opposite page.

Apricot Trees

Apricots appear on the previous season's shoots, but the bulk of the fruit appears on 4-year-old spurs on older wood, and the spurs drop soon after. To encourage spurs, pinch the lateral shoots when they are about 3 inches long.

Heavy pruning is essential to apricot production. Without it apricots will start fruiting with heavy crops, but the crops will dwindle as lush foliage shades the lower fruit spurs. Fruit will then be borne only high in the tree.

APRICOTS

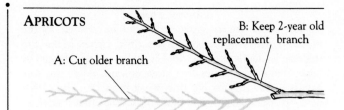

A: Cut older branch

B: Keep 2-year old replacement branch

Plan to shape the young apricot to a wide-spreading head, and keep it low. To maintain this shape, a tree will have to be severely thinned and headed annually. A good apricot tree has a stubby look and no long, thin branches. The rule is to remove one-third of the new wood each year by both thinning and heading. Do this in the winter when the tree is dormant.

When the fruit spurs on a branch are 3 years old—you can tell by counting the annual bud scar rings—select a new branch from lateral shoots on this branch. The next year, remove most of the old branch, cutting it off just above the selected lateral. The fruit will then appear mainly on the new branch. Remove one-third of the older lateral branches each year.

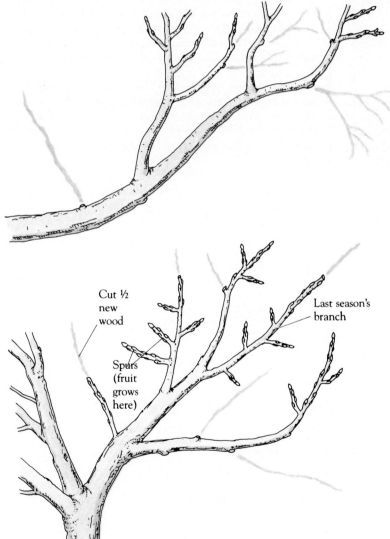

Cut ½ new wood

Spurs (fruit grows here)

Last season's branch

PRUNING APPLES, PEARS, AND EUROPEAN PLUMS

Trim lightly to remove tangled branches or damaged wood. Cut dangling limbs or vertical watersprouts at the base. Head back branch tips to maintain size of older trees. Leave twiggy spurs for fruit production.

HEADING A FORK

To encourage side growth, you can head back a fork, but always cut the two branches to different lengths for a stronger tree.

PRUNING APRICOTS AND JAPANESE PLUMS

Most fruit is formed on short spurs growing on two- or three-year-old wood. Head back new whips by a half. The half you leave will form fruit spurs the following summer and produce a good crop of fruit the year after that. Trim away tangles.

FRUIT GROWTH ON PLUMS AND PEACHES

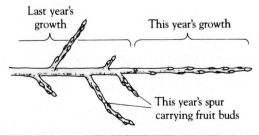

Last year's growth

This year's growth

This year's spur carrying fruit buds

Cherry Trees

As the sweet cherry grows it should be pruned to the modified central-leader system described earlier. Make sure that the leader or upper scaffold branches are not choked by lower scaffold branches that grow upward. After the tree begins to bear, prune out only weak branches, those that develop at odd angles, and those that cross other branches. Be sure to head back the leader and upright side branches to no more than 12 to 15 feet, so that the mature tree can be kept at about 20 feet.

Sour cherry trees differ from the sweet cherries in that they tend to spread wider and are considerably smaller. In fact, some varieties resemble large bushes. The sour cherry can be pruned in a modified central-leader shape, or—if you prefer to keep the tree smaller—prune it to a vase shape. It is quite easy to keep the sour cherry under 12 feet with either system.

Citrus

There really isn't much pruning to do on citrus trees. Pruning should consist of removing any damaged branches (cut off winter-killed branches only after new growth is sturdy), and clipping off dead twiglets inside the tree. Sometimes an extra vigorous shoot suddenly sprouts beyond all the other branches. Reach in and break it off at the base. This will remove other buds that may do the same.

Dwarf citrus may sprout suckers from the rootstock. Break these at the base as soon as you see them.

Fig Trees

Because fig trees bear fruit on wood 1 year old or older, pruning is necessary only to shape a tree for its health and for convenient picking. Prune figs to suit the growing situation. Different varieties grow in different ways. The 'California Black' and the 'Adriatic' grow like a spreading shrub. Do not head these trees back, for they will never grow tall or wide. Select scaffold branches at the first dormant period and prune to keep future branches off the ground. Each year remove low branches that touch the ground or interfere with picking.

The 'Kadota' fig is a vigorous grower that should be kept low and spreading. Head new growth short in the middle of a tree and longer on the outside. When a tree reaches its mature shape, head new growth back 1 or 2 feet.

Peach Trees

Most newly planted young peach trees are pruned to the vase shape. When the tree is planted, cut it off at between 24 and 30 inches in height and leave 3 or 4 laterals to grow into the vase form. These should be headed back the second year only, and only if they exceed 28 to 30 inches in length. These laterals should be spaced evenly around the trunk and 6 to 8 inches apart vertically.

After the second year, peach trees should be pruned moderately. Peaches are borne on the last year's wood, so it is necessary to prune every year to stimulate new growth for the following year's flowers and fruit. The object is to prune for an open center or bowl-shaped tree. Remove all branches other than the main three or four from the trunk and prune off the vertical rising shoots on the remainder. Your ultimate goal is to have a wide tree with an open top 12 or 13 feet high. As the tree grows, cut the branches that are growing upward back to laterals that are growing outward. This creates a wider branching habit.

When the tree reaches 10 to 12 feet and is maturing, start severely cutting back the new growth on the top of the tree being sure to maintain the open center that will admit light to the lower inside parts of the tree. In general, pruning should be lighter on young bearing trees than on older peach trees.

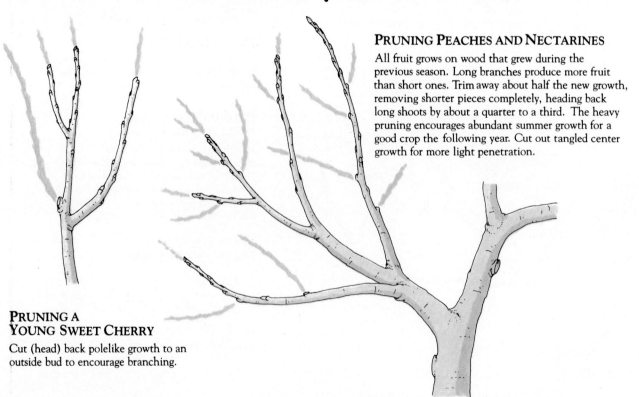

PRUNING PEACHES AND NECTARINES

All fruit grows on wood that grew during the previous season. Long branches produce more fruit than short ones. Trim away about half the new growth, removing shorter pieces completely, heading back long shoots by about a quarter to a third. The heavy pruning encourages abundant summer growth for a good crop the following year. Cut out tangled center growth for more light penetration.

PRUNING A YOUNG SWEET CHERRY

Cut (head) back polelike growth to an outside bud to encourage branching.

Pear Trees

You should train the young pear tree to the modified central-leader system by selecting 5 or 6 scaffold branches over a two-year period. Since it is characteristic of pear varieties to grow upright, be careful not to have too many heading-back cuts, for they will promote too many upright shoots. If you want a small pear tree, buy a dwarf; don't try to make a standard smaller by heavy pruning.

The pear is very susceptible to fireblight, especially in the soft succulent growth that results from heavy pruning, so be careful about heading back or thinning shoots on mature trees. Once fireblight takes hold, there is very little that can be done except to remove infected growth.

Plum Trees

Prune all plums in the winter when they are dormant. Plums fruit on wood produced the previous year and on spurs on older branches. They are particularly prone to branch splitting when mature and bearing heavy crops.

There are many plum varieties, but they fall into two groups—European and Japanese—based on the length of the fruiting spurs. The European spurs may reach 3 feet long, much longer than the 3-inch spur of Japanese types. Since the fruit buds are so spread out, far less thinning is needed and the long, bushlike mass of spurs does not require the severe pruning given Japanese varieties.

Japanese varieties include most of the dessert plums found at the market, such as the 'Santa Rosa' and 'Satsuma'. Some trees grow upright and some spread out; but all fruit in this group is borne on stubby spurs no longer than 3 inches. These spurs will bear for 6 to 8 years.

Remove one-third of the new wood each year by thinning and heading. This heavy pruning is necessary to produce larger fruit. Keep long, thin branches headed to give the tree a stubby, wide shape. When the fruit spurs on a branch have borne for 6 to 8 years, select a new branch from lateral shoots on this branch. The next year, remove most of the old branch, cutting it off just above the selected lateral.

Remember that any heavily pruned tree sends out abundant new growth in the spring. If a tree bears on 1 year or older wood, heavy winter pruning will cut back fruit production the following spring. If it bears on new wood, heavy pruning will stimulate new wood growth and, therefore, will stimulate fruit production the following season.

Pruning always removes some fruiting wood, but an unpruned tree may bear too heavily and may produce small fruit and next to no new growth. Proper pruning produces even crops over many seasons.

Pruning Cane and Bush Berries

Cane berries all fall into two groups: Those with rigid canes that grow upright, and those with trailing canes that tend to creep. Both types produce fruit on canes sprouted the previous year. Cut these canes to the ground as soon as you have harvested the crop. Leave about 5 of the best new ones and cut the rest. Everbearing raspberries differ in that a light crop forms at the tops of new canes in fall. Cut only the portion that fruits; the lower portion will bear the following year.

Rigid-caned berries include blackberry, blackcap raspberry, and purple raspberry. Cut old canes after harvest. Pinch young blackberry canes when they reach 36 inches; pinch blackcaps and purple raspberries at 24 to 30

PRUNING RIGID AND TRAILING BERRIES:

Blackberries and raspberries

Last year's growth is blooming and bearing fruit as new shoots emerge from the crown (those shown near the ground). Remove all but 5 of the new shoots. Let them continue to grow on the ground.

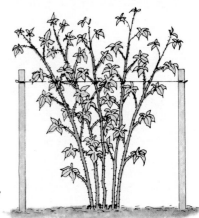

After harvest, cut all of the bearing canes to the ground and tie the 5 new canes to the wire.

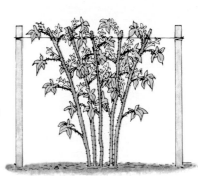

Head back the new canes at a point a few inches above the wire to encourage lateral growth along the wire.

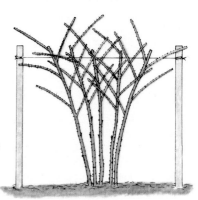

In winter, cut the laterals back to 18″. They will bear next summer and continue the cycle.

TRAINING RIGID BERRIES ON A DOUBLE WIRE

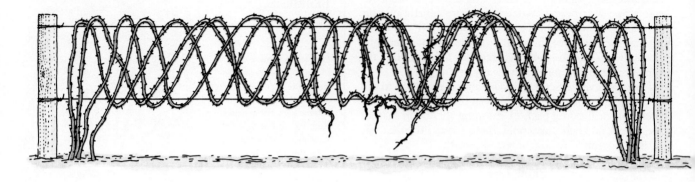

TRAINING TRAILING BERRIES ON A SINGLE WIRE

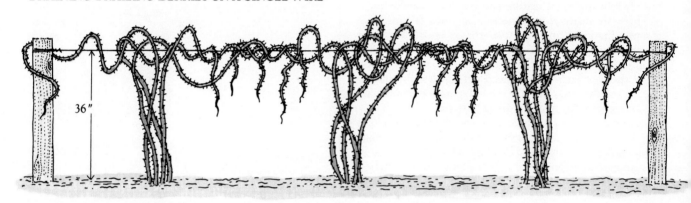

36"

inches. Pinched canes will send out lateral growth. In winter cut blackberry laterals to 15 to 18 inches, and raspberries to 10 to 12 inches. Paint the bases white to distinguish these canes from summer growth. If you wish, tie the erect canes to a wire 18 inches above ground.

Trailing raspberries. For single-crop berries, cut canes as soon as the harvest is over. Train new canes to a post or to one or two horizontal wires. In summer, as new canes grow, gather them in bunches, tie them very loosely, and lay them along the ground until it's time for training. For everbearing raspberries, cut canes that fruit early in the season, and train young canes. These will fruit at the top in fall. Cut the fruiting portion after harvest, but leave the rest.

Trailing blackberries (dewberries) can be treated like raspberries. Another method is to stretch a wire 36 inches above the row of berry plants. After harvest remove old canes and cut off new ones at 48 inches. Canes that sprawl left are then pulled back to the right side of the wire. Canes that grow right are pulled to the left side. Tie if necessary. Extremely long laterals will grow outward, knitting together. Train them as needed along the wire.

Bush berries such as blueberries, currants, and gooseberries tend to bear very heavily if left completely unpruned. You can clean them up by removing the oldest shoots (3 or 4 years old) in winter, thinning out the worst tangles among the twigs, and cutting out dead wood. If berries are very small one year, thin the following winter. If they are large, skip the thinning.

Pruning Grapes

Grapes all require heavy pruning to produce fruit, but after the first three growing seasons, different types of grapes need different pruning. Wine grapes and muscadines usually need spur pruning, in which all side branches of a mature plant are cut to 2 buds in fall or winter. Two new shoots grow on the spur you leave, and each produces a cluster or bunch of fruit.

Some grapes do not produce fruit on shoots that grow too near the main scaffold. 'Thompson Seedless' and many American grapes such as 'Concord' are among these. You must cane prune these grapes. Instead of cutting to a short spur in winter, leave 2 whole canes from the previous growing season. When fruit forms from side growth along this cane, clip the cane off beyond the next set of leaves. You thereby encourage two new canes that will bear fruit the following year. Both spurs and canes grow from a permanent trunk, or trunk-plus-arms (side branches) that you train on a trellis or arbor.

The grape varieties list in the "Encyclopedia of Fruits and Berries" (pages 57–107) indicates whether you should practice spur or cane pruning on a particular variety. In general, all muscadines need spur pruning. Americans of the 'Concord' or fox grape group need cane pruning. Wine grapes usually require spur pruning. For any grape not listed, check with your agricultural extension agent, or experiment by cane pruning a portion of a mature vine, and spur pruning another portion.

PRUNING SPUR AND CANE GRAPES FOR THE FIRST THREE SEASONS

When you plant: Plant a rooted cutting with two or three buds above the soil, then bury those in light mulch. See page 25.

First growing season: Leave the plant alone. It will grow a number of shoots.

First dormant season: Choose the best shoot and cut others to the base. Head remaining shoot to 3 or 4 strong buds.

Second growing season: When new shoots reach about 12" long, select the most vigorous and pinch off others at the trunk. Tie the remaining shoot to a support (arbor post, trellis post). When the shoot reaches branching point at arbor top or trellis wire, pinch it to force branching. Let two strong branches grow, pinch any others at 8 to 10" long.

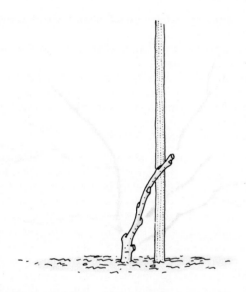

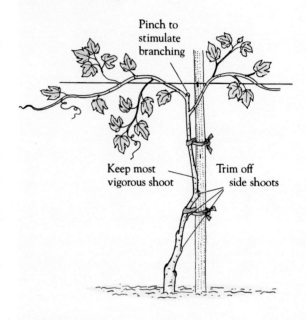

Pinch to stimulate branching

Keep most vigorous shoot

Trim off side shoots

Second dormant season: Cut away side shoots, leaving only the trunk and two major branches. Tie these to the arbor top or the trellis wire.

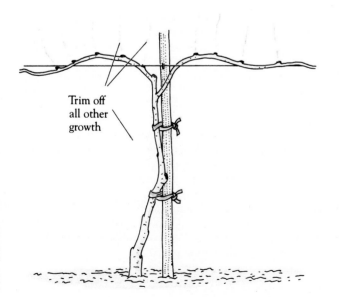

Trim off all other growth

Third growing season: Let the vine grow. Pinch tips of sprouts on trunk. After this, spur and cane pruning differ.

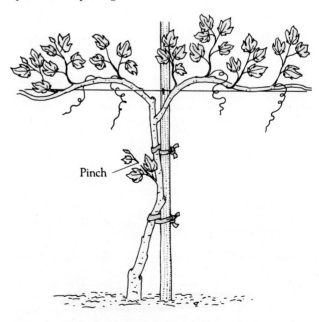

Pinch

TRAINING

Fruit trees can be grown successfully as hedges, garden dividers, boundary plantings, and espaliers. This section details how you can use dwarfing and training to confine fruit trees within tiny spaces.

Home gardeners are not the only ones concerned with limited space planting. Commercial growers are experimenting with training methods that let them grow fruit in hedgerows and harvest their crops without having to haul ladders and climb 30 feet up a standard tree. We combine commercial experience with the classic methods of *espalier* and *cordon* training.

Growing fruit in tight spaces is really no harder than maintaining a healthy rosebush, but keep the following points in mind:

- ☐ Be especially careful about planting and general maintenance
- ☐ Prepare your soil well and, where drainage is a problem, use low, raised beds
- ☐ Feed and water on a regular schedule, and keep a careful watch for signs of any insects or diseases
- ☐ Don't let new growth escape from you and spoil the pattern
- ☐ Inspect your plants frequently

In limited-space planting, training continues all seasons throughout the life of the plant. Be ready to pinch or snip at any time. Major pruning is still a winter task, but in summer you will need to head, or cut away wild growth and suckers, and you may need to loosen or renew ties or add new ones.

Be sure you understand the normal growth patterns of the plants you intend to train. For example, dwarf apples and pears grow slowly and bear fruit in the same places for years. Although fruiting spurs may need to be renewed over the years, the growth pattern means that you can confine these trees to formal shapes and keep them that way. On the other hand, peaches and nectarines fruit on branches that grew the previous year. Old branches will not bear, so they should be cut away like berry canes and replaced with new growth from the base of the tree. This heavy pruning makes rigid training patterns impossible. Peaches and nectarines can be fanned out over walls or grown as hedges, but they cannot be held to strict geometric shapes.

In warm climates you may wish to try training a citrus. You can train them along wires for a fence, or espalier them on walls.

Grapes make good subjects for fences or walls, and a variety that requires a little more heat than your region normally offers may produce good fruit when grown on a south or west wall.

You can train cane berries flat against fences or walls, and treat them something like peaches, since you must replace all canes that have fruited with canes of the current season.

The poorest subjects for limited space training are the quince and cherry. The quince fruits at the tips of new twigs, and the cherry is normally too large to confine and will not fruit at all without a pollinator close by. Both of these plants can be trained, but your efforts would be better spent on something more rewarding.

Top: An arbor covered with lemons.

Bottom: Apples trained along a wire.

Apples and Pears

Both dwarf apples and pears grow and fruit well when trained as hedges against horizontal wires. Use wooden rails in very cold climates. Set posts about 8 feet apart. Stretch a bottom wire between them at 24 inches above the ground. For very small trees, place the upper wire at 4 or 5 feet. For larger dwarfs, place a third wire at 6 or 7 feet.

Plant the young bare-root trees about 3 feet apart, beginning next to an end post. The last tree should be placed about 2 to 3 feet short of the final post. If you buy un-branched trees, bend the trunk at a 45-degree angle and tie it to the wire. If there are any branches with wide crotches, cut them so only two leaf buds remain. Clip off those with narrow crotches at the trunk. Do not feed.

During the first season, train the trunk and any new branches at about 45 degrees, tying loosely where they touch the wires. Pinch off at the tip any branches that seem badly spaced or that extend from the fence at right angles.

The first winter remove badly placed branches at the trunk. Remove the tips from well-placed branches, cutting to a healthy bud on the top of each branch. Feed lightly as growth begins.

The second summer continue training shoots at the ends of branches upward at 45 degrees. Cut side growth to 4 buds beginning in July. Feed again lightly.

Each winter thereafter, remove tangled or damaged growth and cut remaining long shoots to four leaf buds. Feed as growth begins. Each summer, cut out suckers and excessively vigorous sprouts as they appear. Shorten new growth to four leaves from July on, and feed the trees in early August to encourage fruiting wood.

This training method allows side branches to grow outward, away from the fence. Your hedge will eventually become 3 to 4 feet wide. You can hold it at that width by pulling some of the outward growth back toward the fence with string, but check ties frequently or they will cut the branches. If parts of your hedge begin to escape and grow too far outward, trim them back to healthy side branches in May. To maintain the proper height of 5 to 8 feet, cut top growth back to a healthy side shoot in May. Make the cut close to the top wire.

Peaches and Nectarines

Since a peach hedge must have its fruiting wood renewed annually, you will need long replacement branches each year. Plant your hedge as described under apples, using wires at 2, 4, and 6 feet. Cut the whips to about 24 inches long, and shorten those side branches that point along the fence to 2 buds each. Cut off other branches at the trunk. Train all new growth at 45 degrees in both directions. Remove any suckers from below the bud union, cutting to the trunk.

The first winter, cut out about half the new growth at the base, choosing the weakest branches for removal. Cut off the tops of branches you retain if they have grown beyond the hedge limits. Feed lightly as growth begins.

Fruit will form on the branches that grew the previous summer. The original trunk and the lowest branch will form an approximate V shape at or below the lowest wire. During the second summer, choose the healthiest shoots from the lower portions of these main branches, and pinch back all other growth—especially above the second wire—after it

---

TRAINING APPLES AND PEARS

Both apples and pears can be grown as hedges.

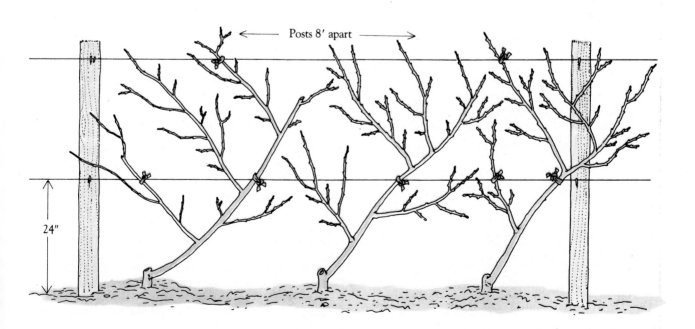

Posts 8' apart

24"

produces 6 to 8 leaves. The lower shoots will replace the entire upper structure and should be tied back loosely to the fence. Continue to remove suckers below the bud union, and feed lightly twice, once during early summer and again in midsummer.

When leaves drop in fall, cut out all branches that have fruited, and head back the V-shaped main structure to the middle wire. Paint all wounds with pruning compound and train the new growth to the fence. Feed as growth begins the following spring. During the summer, again encourage the lower shoots and pinch back the upper growth, feeding twice. Always be sure that there is new growth above the bud union.

This training method will work well against a wall in cool regions where the wall will supply a little extra heat to ripen fruit. Form a wall-trained tree into a fan shape with the outer branches nearly horizontal, the central branches nearly vertical.

Apricots and Plums

Use approximately the same technique described for peaches, but instead of replacing all growth each year, replace about one-third and head back new growth on the remaining branches to 4 to 6 leaves during the summer.

TRAINING PEACHES AND NECTARINES

Each year cut out branches that have fruited, letting new branches replace them to bear next year's fruit.

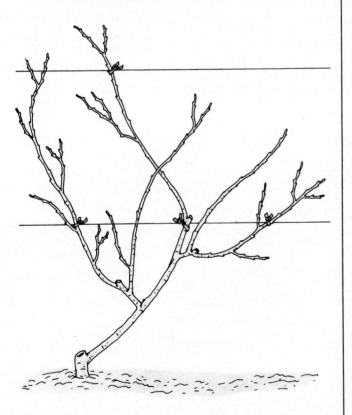

Cane Berries

Canes of the previous season can be trained to a fan or column shape against a wall or used as a fence on a two-wire trellis. (See cane berries under "Pruning," page 43.) New canes should be gathered into loosely tied bundles and placed lengthwise along the wall or fence until old canes fruit. After fruiting, cut out old canes and put the new ones in place. Where disease is a problem, as in many areas of the South, cut and destroy all canes immediately after fruiting and use late-summer growth for the following year's crop.

Grapes

With grape vines you have a choice of arbor training, cordon training, head training, or trellised canes. An *arbor* is an overhead frame on posts at least 8 to 10 feet on a side. Train a vine up each vertical post with two branches crossing the top horizontally. It may take more than two seasons to reach the top and to begin training the horizontals.

A *cordon* is a horizontal permanent branch on a wall or trellis. Train each vine to either 2 or 4 cordons. For 4, allow 3 shoots to grow during the second growing season. Train 3 of the shoots horizontally, tie the other vertically until it reaches the upper support, then pinch and select 2 horizontals.

A head-trained vine is freestanding but may not give much fruit. This method is attractive, and can be used for spur or cane grapes. Stake the young trunk and allow up to 4 shoots to grow, beginning about 24 or more inches above the ground. To spur prune, cut each shoot to 2 buds in winter each year. To cane prune, gather the fruiting canes upward and tie them together toward the tip. Let growth from renewal buds trail.

Trellised canes are grown on wire like a four-arm cordon, but the permanent wood is confined to short stubs near the trunk and the fruiting canes are tied to the wire.

SUMMER PRUNING

Vigorous fruiting varieties will have to be controlled by frequent summer pruning. Summer pruning weakens a plant by removing leaves that manufacture nourishment, and within limited space this pruning is the main means of confining trees. (Winter pruning has the opposite effect, causing a vigorous burst of spring growth.) Too much summer pruning, however, can damage a tree. Experience will teach you how much pruning is necessary, but here are some guidelines.

In early summer, remove only excessively vigorous sprouts that threaten to take over the tree. These may suddenly shoot out much farther than any other growth. Cut them off at the base. Also, remove any suckers from below the bud union, cutting to the base. Paint large wounds with pruning compound.

When the new growth matures and slows its pace, begin snipping it back. The season will vary depending on weather, feeding, and watering, but you can begin to prune some branches in July and finish up by early September. Cut off all but about four leaves of the current season's growth on each new branch. Then give your trees a last feeding of nitrogen to produce new fruiting wood. Don't thin out branches. You can do that during pruning if necessary.

If you find that one tree in a hedge or row regularly overgrows and escapes the pattern you have established, remove this tree and plant a less vigorous one.

SPUR TRAINING OF GRAPES

First three seasons: See page 45.

Third dormant season: Remove all shoots from the vertical trunk. Choose the strongest side shoots on horizontal branches and cut to two buds. Remove weak shoots at the base, spacing a spur, cut to two buds, every 6 to 10".

Annually: Every dormant season after this, each spur will have two shoots that produced fruit during the summer. Cut off weak spurs. Cut the stronger spurs to two or three buds. These buds will produce fruit-bearing shoots in summer. Repeat each year. Always keep the trunk clear of growth. ▶

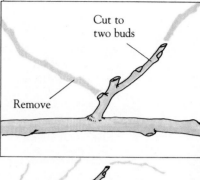

Cut to two buds

Remove

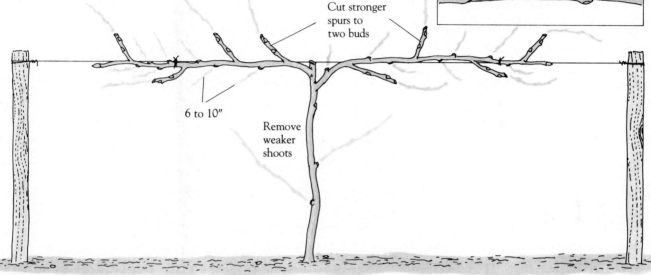

Cut stronger spurs to two buds

6 to 10"

Remove weaker shoots

CANE TRAINING OF GRAPES

First three seasons: See page 43.

Third dormant season: Remove shoots from the trunk. Cut horizontal branches back so that two long shoots remain on each. On a two-wire trellis, you can leave up to eight shoots per vine. Tie the shoot farthest from the trunk to the trellis. Cut the other to 2 or 3 buds. The tied shoot will fruit the following summer. The clipped shoot will produce growth to replace it the next winter, and fruit the year after.

Annually: When the outside cane has borne fruit, cut it back to the inside stub, now holding two or three new canes. Select the best and tie it to the trellis for fruit. Cut the next to two or three buds. Remove the weakest at the base. Repeat each year. ▶

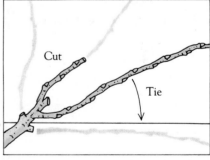

Cut

Tie

Cut to two or three buds

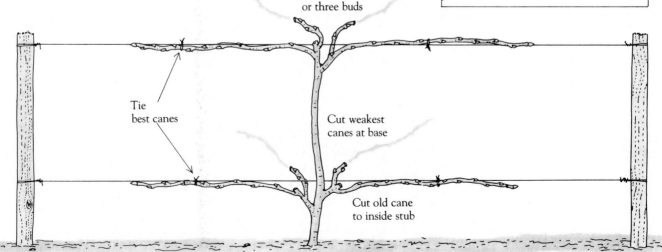

Tie best canes

Cut weakest canes at base

Cut old cane to inside stub

*Using containers, you can plant
a mobile garden and grow a wider
range of fruits and berries than your
climate would otherwise allow.
Specific instructions for watering,
feeding, and repotting container
grown plants.*

FRUIT IN CONTAINERS

At the famed Palace of Versailles in the 1600s, the gardeners of King Louis XIV decorated the grounds with many potted orange trees. During the summer the trees lined the walks of the palace gardens, then when the snows came the gardeners wheeled them indoors to a special greenhouse, which became known as the *orangerie.* Some of these trees are said to have lived for 75 years.

Mobility is the principal reason for growing fruit in containers. Moving fruit plants to shelter when cold weather comes, or to a shady spot if excess heat is the problem, makes it possible to grow them outside their normal climate ranges. With containers you can relocate plants to find out where they do best and even try varieties not usually recommended for your climate, such as lemons in Michigan or peaches in North Dakota. Citrus, in fact, is so attractive in containers that you might consider bringing a small tree into the house for the winter to fill a sunny south window. Deciduous trees can survive a winter season in a garage if given sun on warmer days. One caution, however: Just because a plant can survive winter in the ground in your area does not mean it can manage cold weather in a container. If your garden soil freezes to any depth at all, then container soil is likely to freeze all the way through. Gardeners in the coldest northern zones should plan to protect even hardy deciduous plants during the coldest months of the year.

What fruits can you plant in containers? The answer is:

*Most citrus grow well in containers. Careful
pruning will keep this 'Marsh' grapefruit from
outgrowing its planter for one or two more years.*

Virtually any you like. Trees, of course, must be grafted or genetic dwarfs. The "Encyclopedia of Fruits and Berries," beginning on page 57, lists a great many dwarf varieties of apples, apricots, cherries, nectarines, peaches, pears, and plums, all of which are suitable for container culture. You can also use citrus grafted to trifoliate orange rootstock. Any fig can be grown in a container. Strawberries can be planted in large or small containers, and blueberries and currants make excellent container plants. You can even plant grapes, providing you give them a trellis or other support during the growing season.

CHOOSING CONTAINERS

What kinds of containers are suitable for growing fruit plants? The answer again is: Virtually anything you like, as long as it will hold the plant and a sufficient amount of soil, is nontoxic, and contains holes for adequate drainage. Half-barrels, for example, make excellent containers for fruit trees, as do wooden and ceramic planters. Some people even use rubber waste cans, plastic utility containers available at paint stores, and metal lard cans.

Garden centers, nurseries, and other retailers offer clay, ceramic, porous concrete, plastic, and wooden containers in many sizes and shapes. Plastic and metal containers have the advantage of holding moisture longer. Clay dries out quickly, necessitating frequent watering, and clay pots are quite heavy. Wood has several advantages, not the least of which is its light weight relative to its strength. Containers filled with soil can be extremely heavy; soil alone can weigh about 90 pounds per cubic foot when dry, 100 pounds when

Right: Strawberries and alyssum make an unusual hanging basket. Below: 'Bonanza' dwarf peaches.

wet. Lighter wooden containers can make a significant difference when you're considering relocating containers. Another advantage of wooden containers is that their sides can be screwed together so that they are easy to dismantle, which facilitates the root pruning necessary to keep the plant small. Wood is also readily available to gardeners who wish to make their own containers. See Ortho's book, *Wood Projects for the Garden.*

Whatever type of container you choose, the size should be just 2 or 3 inches wider than the roots of your plants. The right size container will let the plant find water and nutrients easily, keep soil from going sour around and beneath the roots, and slow top growth. If you start with a bare-root apple, a pear, or one of the genetic dwarf fruits, your first container should be about the size of a 5-gallon can. Let the young tree grow for a season and then repot it the following spring in a larger container. Move plants from the nursery 5-gallon-size container to the bushel size over two or three seasons.

Evergreen fruit plants such as citrus should be started in a container not much bigger than the rootball. If your soil mix has good drainage you can use a box 3 or 4 inches wider than the roots all around. For large nursery plants—for example, those in 15-gallon cans—the first container may be a permanent one. To hide the metal can, you can place it in a large basket or surround it with a box fashioned from wood panels.

If you plan to relocate the plant and its container, carefully consider container size in advance. The maximum for permanent containers should be about bushel-basket size. Anything bigger will be too bulky to handle or move. Half-barrels, or any boxes or pots that hold about an equal volume of soil, are roughly the right size. The minimum permanent size should be about 18 inches on a side and 18 inches deep. The smaller the container, the easier it is to move; however, keep in mind that the plant must have sufficient room for its roots and that more work is involved in feeding, watering, and root pruning with smaller containers. All these factors must be balanced when choosing the best container.

CONTAINER SOILS

Because containerized soil loses moisture easily, the soil must hold moisture well. To prevent soggy roots and the possibility of disease, it must also have good drainage. Commercial mixes such as Jiffy-Mix and Redi-Earth, often referred to as "soilless mixes" or "synthetic soils," fit the bill. Jiffy-Mix, the most widely available synthetic soil, is made of 50 percent peat moss and 50 percent vermiculite. It contains enough nutrients to sustain initial plant growth. It also provides fast drainage and a reservoir of air and water after drainage.

Synthetic mixes offer several advantages. They are free of disease organisms, insects, and weed seeds. They are lightweight—half the weight of garden soil when both are wet—an advantage both in relocating container plants and in growing them on roofs or balconies. In addition, they can be used just as they come from the bag, without needing to be moistened for planting.

If you plan to fill a number of containers, you may want to save money by mixing your own planting medium. Here is a basic recipe:

9 cubic feet of fine sand
9 cubic feet of ground bark

Add to the above:

5 pounds 5–10–10 fertilizer
5 pounds ground limestone
1 pound iron sulphate

Some gardeners like to add a little rich loam to the mix of sand and organic material. Add up to one-third loam if you like, but be careful not to include clay soil, which holds too much water for a container mix. Also remember that if you add topsoil to the mix when planting in containers, you may give the mix good physical properties but you will also increase the risk of introducing soil pests and diseases that can harm your plants.

POTTING AND REPOTTING

Before you pot up your fruit plant, make sure your container has drainage holes. Cover the holes with broken pieces of pot, glass, or crockery, but don't cover the holes tightly or you'll retard drainage. Do not fill the bottom with rocks or coarse gravel; these do not improve drainage.

To plant a bare-root plant, place enough tamped down soil mix in the bottom of the pot so that the plant crown is slightly below the container rim when the roots are touch-

These unusual containers have built-in trellises for grapes.

STEPS IN REPOTTING

1. Remove plant in early spring. Shave 1″ from sides of roots 1½″ from the bottom, using a large knife.

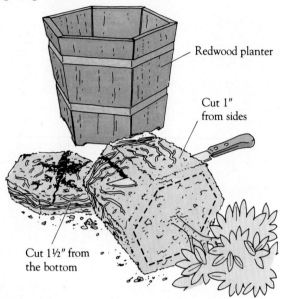

Redwood planter

Cut 1″ from sides

Cut 1½″ from the bottom

2. Reset the plant on fresh soil with the crown just below the rim of the container.

Set the base of the trunk 1″ below the container rim

Fresh soil

3. Fill in around sides, tamping lightly with a stick as you go.

4. Water slowly until water flows freely from the bottom. Prune lightly, removing tangles and extra long branches.

ing the mix. Hold the plant at that level and toss in enough mix to support it, tamping lightly as you go, and filling the container to about an inch below the rim. The soil will settle, leaving room to water. Plants removed from nursery containers can be placed on the first layer of soil. With these you simply fill in soil mix around the rootball; but first scratch the rootball all around with a fork to rough up roots and direct them outward. Be sure to cut off long spiraling roots at the bottom of the rootball.

Repotting is similar, as shown in the illustrations on this page. Repotting is necessary because plants tend to bunch feeder roots at the wall of a container, where they dry out faster, creating shortages of water and nutrients even when you are providing proper care. When you shave off an inch of root and add fresh soil, the plant will grow healthy young roots in the new reservoir of moisture and nutrients. Be sure to clip back the top of the plant when you shave its roots in order to provide a balance between the two. New top growth will soon follow new root growth. After potting or repotting, give the plant a good, deep watering.

FEEDING CONTAINER FRUIT PLANTS

Use the growth and appearance of the plant as guides to feeding. The plant should leaf out and grow vigorously in the spring and early summer, and leaves should be a healthy medium green. Yellowed leaves suggest a lack of nitrogen, while very dark leaves may indicate overfeeding.

If you are using a purely synthetic mix, you must be careful about feeding. The nutrients you add often wash through the soil when you water, so you'll have to feed more often. It's best to keep to a regular schedule.

One good feeding method is to give each plant about one-half the recommended quantity of complete fertilizer (one that contains nitrogen, phosphorus, and potassium) every 2 or 3 weeks. A liquid fertilizer is easier to measure in exact proportions and is also less likely to burn roots. If the label recommends 1 tablespoon per gallon of water each month, use 1½ teaspoons instead and feed every 2 weeks.

Another good method is to use slow-release fertilizer pellets. These dissolve slowly over a period of time, releasing nutrients with every watering.

Feed from the beginning to the end of the first growing season if the plant is to receive winter protection. Stop about mid-July if the plant is to stay outdoors. This will give it a chance to harden new growth.

Citrus

Citrus trees require about the same amount of feeding as deciduous fruit trees, but they may also require a few extra nutrients. Special citrus foods containing iron, zinc, and sometimes other minerals are available at nurseries. Use them regularly, or switch to them if you see leaves with yellowed portions between bright green veins; this may be an iron deficiency. If the leaf is uniformly yellow, veins and all, the plant lacks sufficient nitrogen. Citrus food will not hurt deciduous plants if you want to use it, but it is a more expensive fertilizer.

WATERING CONTAINER PLANTS

Judge when to water by the behavior of your plant. It should never wilt, but it shouldn't stand in soggy soil either. If you check the soil occasionally by digging down an inch or two, you'll soon learn when and how much to water. The top inch may stay moist for a week in fairly cool weather, but in hot, windy weather you'll need to water more often, perhaps even every day for a plant that needs repotting. (This is why well-drained soil is important: You can water liberally without drowning the roots.)

Don't count on rain to do all of your watering. The foliage of plants in containers can act as an umbrella, shedding most of the rainfall. Check the soil even when rain has been abundant.

Vacation Watering

When you leave home for a long period, group your containers near a water source and away from the afternoon sun. Grouping them will help keep them moist, shade will further cut the need for water, and if they are located near a hose, your vacation waterer won't miss any of them by accident. For large numbers of containers, you can hook up a permanent system of small hoses and add a timer that turns water on at regular intervals. Drip systems are particularly effective, provided you filter the water before it goes into the system.

'Meyer' dwarf lemon is particularly suited to growing in a container.

Leaching

It is important to leach container soil occasionally to remove built-up mineral salts that can burn leaves. Salts accumulate from fertilizers and from hard water. (Any water that won't produce good soapsuds or leaves bathtub rings is hard water and has a high salt content.) You'll know you have a salt problem when you see brown leaf edges. Leaching is a means of running enough water through the soil to wash away the harmful salt buildup. Avoid disaster by putting your garden hose in each container every couple of months and letting it run slowly for about 20 minutes. The water should flow just fast enough that it soaks through the soil thoroughly and out the drainage holes of the container. It is a good idea to fill the container until water runs freely from the bottom, go on to other containers, then return and repeat the process. This will keep salts to a minimum. You should also avoid using softened water on your plants because it contains harmful chemicals.

MULCH

Mulch will help keep the soil in your containers moist and cool. Use a coarse organic mulch such as bark chips and pile it about 2 inches thick.

This chapter is a cornucopia of photographs with flavor and color descriptions for hundreds of fruit and berry varieties. Varieties are listed according to hardiness, time of fruiting, and special care requirements.

ENCYCLOPEDIA
OF
FRUITS AND BERRIES

The information in this encyclopedia is designed to help you choose the best possible fruits for your garden. Each entry contains notes on tree or plant size, planting, how the fruit grows, pruning and thinning, pollination, and pests and diseases, and, where applicable, special tips on cultivation.

This is followed by variety lists that contain many of the best and most popular fruit varieties, selected to provide good choices no matter where in the country you live. Some varieties are known by more than one name. In these cases the most common name is used with the lesser known alternatives following in parentheses. Some varieties are grouped according to the time they are harvested relative to the growing season; for example, "Early," "Midseason," or "Late." Others are grouped by type of bearing, growth habit, or hardiness. You'll find information on where the variety originated (which may give you a valuable clue about how well it will do in your area), a description of the fruit and the plant, special growing requirements and other distinctive characteristics, and the best uses for the fruit once it's harvested. In addition, where possible we tell you in which section or sections of the country the variety performs best.

Bear in mind that just because a fruit is recommended for one area does not necessarily mean that it can't do well in others. Climatic conditions vary so widely that any such recommendations must be general. Local geographical features and other factors that affect climate, as well as special treatment from the gardener (for example, providing sheltered areas or winter protection), can support plants that generally are not expected to do well in a given region. If you are attracted to a particular variety but are uncertain whether you can grow it in your garden, check with your local nursery staff, county extension agent, or gardening neighbors.

Many varieties have chilling requirements that must be met for the fruit to develop properly. The chilling requirements are described in terms of the number of hours of exposure to winter temperatures below 45°F. Fruits and berries that require chilling fall into one of three general categories:

☐ Low-chill   300 up to   400 hours below 45°F
☐ Moderate-chill 400 up to   700 hours below 45°F
☐ High-chill   700 up to 1000 hours below 45°F

When a variety is described as low-chill, moderate-chill, or high-chill in this encyclopedia, use the above guidelines unless more specific numbers are given for a particular variety.

If unusually warm weather occurs in the winter, the rule of thumb is that for every hour over 70°F that interrupts the chilling process, you lose one hour of chilling that has already taken place. A warm winter can prevent fruit trees with a high-chill requirement from setting a good crop.

A list of catalog sources is on pages 107 and 108. You can order directly from many of these dealers. If the source is a wholesaler, you should order the varieties through your local nursery, who in turn should be able to get the plants you want from the wholesaler.

Ripe 'Royal Ann' cherries are ready to be picked. This is the variety that is colored and flavored to make maraschino cherries.

Right: Standard apple tree
*Below: An informal
espaliered apple*

APPLES

For some people, the only apple is a polished red one. Others must have a red-and-yellow apple, or a golden one, or a tart green apple for pies. To try to cover these preferences, the variety list contains a broad selection of the most popular apples for the home gardener. You'll find the popular varieties, 'Delicious', 'Jonathan', and 'McIntosh', and many of their sports. You'll also find the newest hybrids developed for special conditions and purposes.

There are well over a thousand apples available today. Many of these have been developed through the painstaking, time-consuming efforts of apple breeders. In breeding, each parent plant supplies half the heritage of seedlings, but that half may be a set of characteristics either partly or completely hidden in the parent. The seedlings are an unknown mixture until breeders grow them to fruiting size to see what characteristics they do have. This work takes time, and many seedlings prove to be inferior to their parent trees.

Sports, or mutations, may occur at any time, often without apparent reason: Suddenly one branch of a tree is different. Occasionally the odd branch results from mechanical damage, such as pruning; sometimes experimenters purposely change genetic structure with chemicals or radiation. Most sports are worthless, but now and then one turns out to have characteristics that make it worth propagating to create a new strain.

One way to arrive at a general understanding of apple breeding is to look at the "heads of families." Three of these are 'Delicious', 'Jonathan', and 'McIntosh'.

Top: Apple blossoms
Center: Informal espalier
Bottom: Oblique cordon espalier
Below: A "spur" type tree

'Delicious', which is by far the most popular and economically important apple in America, first sprouted in an Iowa orchard in 1872. Its parentage is uncertain, but one parent may have been a nearby 'Yellow Bellflower' apple. That 'Delicious' exists at all today is almost a miracle. The owner, Jesse Hiatt, cut the seedling down twice, but it resprouted, so finally he let it grow. In about 1880 it bore fruit, which Hiatt thought was the best he'd ever tasted. The name 'Delicious' was given at a fruit show by C.M. Stark of Stark Nurseries. Stark didn't learn the name of the grower until 1894, but by then the apple had begun its rise to fame. 'Delicious' has produced a number of sports, including the original red sport, 'Starking'; the redder 'Richard', 'Royal Red', 'Hi Early', 'Chelan Red', and 'Red Queen'; and the spur-type 'Starkrimson', 'Redspur', 'Wellspur', 'Hardispur', and 'Oregon Spur'. 'Delicious' is also a parent of 'Melrose'.

The first seedling of 'Jonathan' sprouted in Kingston, New York, apparently from the fruit of an 'Esopus Spitzenburg.' A Judge Buel of Albany found the apple so good that he presented specimens to the Massachusetts Horticultural Society, naming it for the man who first showed it to him. 'Jonathan' was the primary variety before 'Delicious' took over.

Red sports of 'Jonathan' include 'Jon-A-Red' and 'Jonnee'. Hybrid descendants include 'Jonagold', 'Jonamac', 'Idared', 'Melrose', 'Minjon', and 'Monroe'.

The 'McIntosh' apple came from the McIntosh Nursery in Ontario, Canada. John McIntosh discovered it about 1811, but did not propagate grafted stock until 1835, when the grafting technique was perfected. Well-known descendants of 'McIntosh' include 'Summerred', 'Niagara',

'Early McIntosh', 'Puritan', 'Tydeman's Red', 'Jonamac', 'Macoun', 'Empire', 'Cortland', 'Spartan', and the spur variety, 'Macspur'.

Other apples with long lines of descendants include 'Rome', 'Golden Delicious', 'Northern Spy', and 'Winesap'.

The extensive work on dwarfing rootstocks for apples has produced plant sizes ranging from a 4-foot bush to a 30-foot spreading tree. There is even a true, or genetic, dwarf that stays small on any rootstock.

Dwarfing rootstocks make it easy for a gardener to control size even further by the special techniques outlined in this book. See pages 7–8 for details on dwarfing fruit trees through the various methods of pruning, girdling, scoring, and training. Apples can now grow in boxes, flat on a wall or trellis, as hedges, or in fanciful three-dimensional shapes.

Apples bear on long-lived spurs, so only very heavy pruning will remove your crop. Any training method in this book will suit any apple. See pages 35–48 for specific instructions on pruning and training.

The fruit on an apple forms at the tip of last year's spur growth, and the spur itself then grows a bit more, off to the side of the fruit. Each spur bears for 10 years or more, so don't tear it off when you pick.

Spur-type varieties are sports of standard varieties. They grow more slowly than other plants, and their spurs are packed closer together on the branch. This less vigorous growth means that they are a kind of genetic dwarf, but they are still good-size trees unless grafted to dwarfing roots. Spur varieties are difficult to train formally. If you buy spur varieties on dwarfing roots, use a

Below: A whimsical espalier

training method that doesn't call for any particular form.

Pruning methods depend on how you grow the tree. For special training, turn to pages 46–47. For general pruning of the larger dwarfed trees or standard trees, see pages 40–41. To prune apple trees on Malling 9 dwarfing rootstocks, use the spindle-bush method illustrated on page 40.

Thinning is crucial with many apple varieties. If left alone, the trees set too many fruits and the heavy crop can snap branches. Even more important, many apple varieties tend to bear every other year. If you leave too much fruit you encourage this alternate bearing: The following year you may find that your tree bears only a handful of apples because the large crop of the previous year has depleted the tree's reserves.

There are many thinning methods, but the most direct is to wait for the natural drop of young fruit in June, then thin the remaining fruit so that there is a single apple every 6 inches along the branches. Each spur may have a cluster of fruit. A single fruit is less likely to become diseased, so leave only the largest fruit on each spur. Thin carefully or you will damage the spurs.

Apples are only partially self-fertile, but many varieties set a good crop without a pollinator. Any two kinds that bloom together offer cross-pollination except for 'Gravenstein', 'Winesap', 'Stayman', and 'Stayman' sports such as 'Blaxstayman' and 'Staymared'. These will not set a crop at all if you plant them with no other source of pollen. Also, if you plant only a very early and a very late variety, they will not cross-pollinate. A few apples are completely self-sterile.

All apples need some cool winter weather, but there is an enormous range in this requirement, so varieties are available for any climate except subtropical and low desert regions.

Apples are subject to attack by many organisms, but the gardener will have most trouble with codling moth and other fruit-spoiling pests and then with the usual aphids, mites, and scales. See the pest and disease section on page 29–33 for further details. A regular spray schedule is best. Repeated sprays can control diseases such as mildew and scab.

Early Season Varieties

◀ **Lodi.** Origin: New York. The fruit is up to 3 inches in diameter, with light green skin, sometimes with a slight orange blush. The flesh is nearly white with a greenish tinge, fine-grained, tender, and juicy, but sour. The eating quality is only fair, but 'Lodi' is excellent in sauce and pies. The tree tends to overset fruit and must be thinned early. Widely available.

◀ **'Jerseymac'.** Origin: New Jersey. A McIntosh cross that ripens in August. The red fruit is medium firm, juicy, and of good quality. The tree produces a crop every year and is generally available.

◀ **'Tydeman's Early'.** Origin: England. A 'McIntosh' type, similar in shape and ripening four weeks earlier, this apple is almost entirely red from a very early stage. Fruits should all be picked within a few days for optimum quality and flavor, and because they drop quickly at maturity. They are good eating quality and keep much longer than most early varieties. Early ripening, when few other varieties are being harvested, is a virtue. One drawback is growth habit: The branches are undesirably long and lanky, and need to be controlled by pruning. For best results, grow this one on dwarf or semidwarf rootstocks. Widely available.

◀ **'Yellow Transparent'.** Introduced from Russia almost 100 years ago, this apple is still valued by a loyal following as an early green cooking apple. It is medium-size with greenish-yellow skin. The texture is fine grained, crisp, and juicy. It bruises readily and soon becomes overmature. It is good fresh and excellent for sauce and pies. Widely available.

Below: 'McIntosh'

Early to Midseason Varieties

◀ **'Gravenstein'.** The fruit is large but not uniform, with skin that's red against light green. The greenish yellow flesh is moderately fine textured, crisp, firm, and juicy. It is excellent for eating fresh, in sauce, and in pies. Trees are strong, very vigorous, upright, and spreading. Widely available along with 'Red Gravenstein'.

◀ **'Jonamac'.** This 'McIntosh' type dessert apple is of very good eating quality, milder in flavor than the 'McIntosh'.

◀ **'McIntosh'.** If you write down the attributes of a great apple—medium to large fruit with sweet, tender, juicy, white flesh, very good fresh or in sauce, pies, or cider—you are describing 'McIntosh'. The skin is yellow with a bright red blush. The tree is strong and very vigorous. Widely available.

◀ **'Paulared'.** Origin: Michigan. This apple rates high on several counts. It has an attractive solid red blush with a bright yellow ground color. The flesh is white to cream and nonbrowning. Its excellent, slightly tart flavor makes it good for eating fresh and in sauce and pies. Although it colors early, for quality apples it should not be picked until nearly mature. Fruit holds well on the tree and is harvested in two pickings; it has a long storage life. The tree is everything an attractive tree should be—strong and upright, with good branch structure.

◀ **'Prima'.** Origin: Illinois. This red juicy apple has fair quality, but its main feature is its resistance to scab, mildew, and fireblight.

◀ **'Wealthy'.** This hardy old-timer is good for both eating and cooking. A long bloom period in midseason makes it a good pollinator for most other varieties but this apple requires heavy thinning. Fruit is medium to large, uniform, and rough. Flesh is white stained with pink, fine textured, firm, tender, tart, and juicy. It is good fresh and for pies, baking, and stewing, and it is excellent for sauce. Widely available.

Midseason Varieties

◀ **'Cortland'.** Origin: New York. According to many apple growers, this is excellent—even better than 'McIntosh' —as a dual-purpose apple, for eating and cooking. The tree bears heavy crops of large red-striped fruit with white flesh, which is slow to turn brown when exposed to air, making it especially suited for use in salads. The tree is strong and very vigorous, with a spreading, drooping growth habit. Widely available.

◀ **'Empire'.** Origin: New York. This cross between 'McIntosh' and 'Delicious' has medium, uniform fruit with dark red, striped skin and whitish cream flesh that is firm, medium textured, crisp, and very juicy. Eating quality is excellent. A major fault is that it develops full color long before maturity, tempting the grower to harvest too early. Trees are moderately vigorous and of upright-spreading form.

◀ **'Jonathan'.** Origin: New York. The standard 'Jonathan' is one of the top varieties grown in commercial orchards in the Central States. The fruit is medium size and uniform; the skin is washed red and pale yellow; and the flesh is firm, crisp, and juicy. Rich flavor makes it a good choice for snacks, salads, and all culinary uses. Trees bear heavily. Widely available.

Below: 'Spartan'

Below: 'Golden Delicious'

◀ **'Macoun'.** Origin: New York. There was a new burst of interest in this variety in the early 1970s because of its excellent quality when eaten fresh. A cross of 'McIntosh' and 'Jersey Black', it resembles 'McIntosh' but is smaller. Skin is very dark red in color; flesh is white, richly flavored, aromatic, and of high quality as a dessert fruit. The tree grows upright with long, lanky branches. Thinning aids in attaining good fruit size. Widely available.

◀ **'Rhode Island Greening'.** Still rated at or near the top as a cooking or processing variety after more than 200 years, this apple is light green to yellow, firm fleshed, crisp, and juicy. It is a top-quality choice for sauce and baking. The tree generally produces good crops but is a poor pollinator. It bears in alternate years.

◀ **'Spartan'.** Origin: British Columbia. This is a cross between 'McIntosh' and 'Yellow Newtown'. The fruit is medium size, uniform, and symmetrical, with a solid dark red skin. Flesh is light yellow, firm, tender, crisp, and juicy. The tree is strong, moderately vigorous, and well shaped. It must be thinned to assure good size and annual bearing. Widely available.

Midseason to Late Varieties

◀ **'Delicious'.** The most important apple grown in the United States, the apple is medium size, long, and tapering. The skin is striped-to-solid red with slightly yellow flesh that is firm, juicy, sweet, and aromatic. Hand thinning is usually necessary to produce apples of good size and dessert quality.

◀ **'Golden Delicious'.** For a great eating and cooking apple, 'Golden Delicious' ranks as high as any. The fruit is medium to large and uniform in size. The skin is greenish-yellow with a bright pink blush. The flesh is firm, crisp, juicy, and sweet —excellent fresh and in desserts and salads, and very good for sauce. The tree is of medium height, moderately vigorous, upright, and round, with wide-angled crotches. It bears very young and continues to bear annually if thinned. This is an excellent pollinator and will set some crop without cross-pollination. Widely available.

◀ **'Jonagold'.** Origin: New York. A cross of 'Jonathan' and 'Golden Delicious', this is a beautiful large apple with a lively yellow-green ground color and bright red blushes. The cream-colored flesh is crisp and juicy, with good flavor. It is good for cooking and among the very best apples for fresh eating. Stores well. The trees are vigorous and sturdy, with wide-angled branches.

◀ **'Red Delicious'.** This is the number one supermarket apple. There is no question about its dessert and fresh-eating quality. The fruit is medium to large in size, with skin color striped to full red. The flesh is moderately firm in texture and very sweet and juicy. Your best choices are the red sports such as 'Wellspur' or 'Royal Red'. The tree tends to produce full crops every other year unless properly thinned for annual bearing. Widely available.

◀ **'Yellow Newtown'.** The fruit is of medium size with greenish yellow skin and crisp, firm flesh. It is good for eating fresh, and excellent for sauce and pies. The trees are strong and vigorous. Widely available.

low: 'Granny Smith'

Center, top to bottom:
'Mutsu', 'Rome Beauty',
'Northern Spy'

Below: 'Idared'

Late Varieties

◀ **'Granny Smith'**. This apple is from Australia and New Zealand and is sold in supermarkets here. The fruit is medium to large and bright glossy green. The flesh resembles 'Golden Delicious', but is tarter. It is very good eaten fresh or in desserts, salads, sauce, and pies. The tree is strong, vigorous, upright, and spreading; but it can only be grown in areas with a very long growing season. Widely available.

◀ **'Idared'**. A cross of 'Jonathan' and 'Wagener', this hybrid has an attractive, nearly solid red skin with a smooth finish. Of large, uniform size, it has white, firm, smooth-textured flesh that is excellent for eating fresh and for cooking. It has a long storage life. The tree is strong, vigorous, upright, and very productive. Widely available.

◀ **'Mutsu'**. Origin: Japan. A cross of 'Golden Delicious' and the Japanese 'Indo', this relative newcomer has gained the approval of both growers and consumers. Large, oblong, greenish fruits develop some yellow color when mature. The flesh is coarse, firm, and crisp. The flavor is excellent (tarter than 'Golden Delicious') when eaten fresh, and it is good for sauce, pies, and baking. Unlike 'Golden Delicious' it does not shrivel in storage. The tree is very vigorous and spreading. Widely available.

◀ **'Northern Spy'**. Trees of this variety are very slow to begin bearing; sometimes 14 years elapse before they produce their first bushel. The fruit is large, with yellow and red stripes, and the flesh is yellowish, firm, and crisp. The quality is excellent fresh and for pies. The fruit bruises easily, but

has a long storage life. Trees are vigorous and bear in alternate years. Widely available.

◀ **'Rome Beauty'**. Origin: Ohio. This variety and its sports are the world's best baking apples. Many red sports (such as 'Red Rome') are available with a beautiful, solid, medium-dark-red color. The fruit is large and round, and the flesh is medium in texture, firm, and crisp. The tree is moderately vigorous, starts to produce at an early age, and is a heavy producer. The fruit has a long storage life. Widely available.

◀ **'Stayman'**. Origin: Kansas. This variety is a very late ripener. Where it can be grown, it is good for cooking or eating fresh. The fruit is juicy with a moderately tart, rich, winelike flavor. The skin is bright red and the flesh is fine in texture, firm, and crisp. Skin cracking is one drawback. The tree is medium size and moderately vigorous. Widely available.

Extra Hardy Varieties

In cold-winter areas where some of the favorite apple varieties are subject to winter damage, gardeners may choose one of three hardy varieties developed by the University of Minnesota.

◀ **'Honeygold'**. Midseason to Late. This apple boasts a 'Golden Delicious' flavor. The fruit is medium to large with golden to yellow-green skin and yellow flesh that is crisp, smooth, tender, and juicy. It is good for eating fresh and in sauce and pies. The tree is moderately vigorous.

◀ **'Red Baron'**. Midseason. This cross of 'Golden Delicious' and 'Red Duchess' has round, medium-size fruit with cherry red skin. The flesh is crisp and juicy with a pleasantly tart flavor. It is good eaten fresh or in sauce and pies.

New varieties of apricots are hardy in most of the United States. This commercial orchard is in California

Below: This tree ripened apricot is ready to be picked.

◄ **'Regent'.** Late. This cultivar is recommended for a long-keeping red winter apple. The fruit is medium size, with bright red skin and creamy white, juicy flesh of crisp texture. Rated excellent for cooking or eating fresh, it retains its fine dessert quality late into winter. The tree is vigorous.

Low Chilling Varieties
◄ **'Anna'.** Early. This apple from Israel flowers and fruits in Florida and southern California. The apple is green with a red blush and fair quality. It is normally harvested in July but sometimes sets another late bloom that produces apples for the fall. Use an early blooming variety as a pollinator.

◄ **'Beverly Hills'.** Early. This is a small- to medium-size apple, striped or splashed with red over a pale yellow skin. The flesh is tender, juicy, and tart. Overall, the apple resembles 'McIntosh'. Use it fresh or cook it in sauce or in pies. The tree is suited mainly to cooler coastal areas, since heat spoils the fruit. Locally available.

◄ **'Winter Banana'.** Midseason. The large fruit is strikingly beautiful. The skin color is pale and waxy with a spreading pink blush. The flesh is tender, with a wonderful aroma and tangy flavor. 'Winter Banana' requires a pollinator such as 'Red Astrachan' in order to set a good crop. Locally available.

◄ **'Winter Pearmain'.** Midseason. A large green apple with moderately firm flesh and fruit of excellent quality. It is a consistent producer in southern California.

APRICOTS
The selection of apricot varieties is limited in the colder regions of the country, because apricots bloom early and may suffer frost damage. In recent years, however, breeders have produced a number of hybrids with hardy Manchurian apricots, and now varieties such as 'Moongold' and 'Sungold' will fruit fairly regularly even in the northern plains. The choice of varieties widens in milder regions, and more tender varieties, such as 'Moorpark', will bear even in the eastern states.

Dwarfed apricots on special rootstocks produce fair-sized trees, and a full-sized tree will fill a 25-foot-square site; but you can train the tree to branch high and use it in the landscape as a shade tree.

Apricots, like plums, bear on spurs that produce for 2 to 4 years and then need to be pruned out and replaced with younger wood. See the sketch on page 41. Fruit may form in the second year, but don't expect a heavy crop until the third or fourth year. Trees are fairly long-lived and may last from 15 to 30 years, depending on location and care.

Many apricots are self-fertile, but in colder regions it is usually best to plant a second variety for pollination to encourage the heaviest fruit set possible. Frost damage may remove many of the young fruits.

In pruning apricots, you need to head back long new whips by one-half and remove the oldest fruiting wood. Thinning is generally natural, either from frost or from natural drop in early summer. If your tree sets heavily, you will get larger apricots by thinning to 2 inches between each fruit. For pruning details, see page 41.

Apricots can also be used as stock plants for grafts. Plums do well on apricot stock, and peaches may take, although the union is weak. Your apricot tree can bear several different fruits over a long season when you mix grafts.

Brown rot and bacterial canker are serious pests.

Varieties

Check for climate adaptability and pollinating requirements, and be sure to buy hardy trees in the colder regions.

◄ **'Earligold'.** Origin: California. This is a medium-size fruit with rich golden skin and juicy flesh. The tree is a heavy producer and requires little winter chill. Ripening begins in mid-May. Good for Southern California. Locally available.

◄ **'Early Golden'.** This medium to large, pale yellow fruit is blushed red, with smooth skin and fine flavor. It ripens in mid-August in New York. Good for the North and South.

◄ **'Hardy Iowa'.** Origin: Iowa. The fruit is pale yellow and rather small, with thin skin and very sweet flesh. Good eaten fresh or in pies and preserves. The tree is a prolific bearer and flowers late, escaping frost. Good for the North.

◄ **'Moongold'.** Origin: Minnesota. A hybrid with 'Manchu' as one parent. The same cross also produced 'Sungold', and the two must be planted together for pollination. The fruit is orange with tough skin. The flesh is orange-yellow and of very good quality. The tree is a spreading, medium-size plant. Fruit ripens in Minnesota in late July before 'Sungold'. Good for all zones.

◄ **'Moorpark'.** Origin: England. This variety, dating from 1760, is considered by many to be the standard of excellence among apricots. The large fruit is orange with a deep blush, sometimes overlaid with dots of brown and red. The flesh is orange, of excellent flavor, and has a pronounced and agreeable perfume. Ripening is uneven, with half the fruit still green when the first half is already ripe. This is

an advantage in the home garden, since the gardener does not have to use the fruit all at once. The tree does well in all but the most extreme climates. Widely available.

◄ **'Perfection' ('Goldbeck').** Origin: Washington State. The fruit is very large, oval and blocky, and light orange-yellow without a blush. The flesh is bright orange and of fair quality. The tree is vigorous and hardy, but blooms early and so is uncertain in late-frost areas. Since it requires little winter chill it will grow in mild-winter areas. It needs a separate pollinator and sets a light crop. In the Southwest use another early variety. Good for the South and West.

◄ **'Riland'.** Origin: Washington State. The fruit is large and rather flat. The light yellow skin is covered over half the fruit by a deep blush. The flavor is rich and plumlike, but the texture is somewhat coarse. The tree is vigorous and upright and requires a pollinator. Good for the West.

◄ **'Royal' ('Blenheim').** Origin: England. The fruit is medium size and flat orange with some tendency to have green shoulders. It is the best drying, canning, and fresh eating apricot in California. It requires moderate chilling and will not tolerate excessive heat (over 90°F) at harvest time.

◄ **'Scout'.** Origin: Manitoba. This variety originally came from a Manchurian fruit experiment station. The flat, bronzy fruit is medium to large with deep yellow flesh. It is good fresh and can be canned or used in jams. The tree is tall, upright, vigorous, and hardy. Fruit ripens from late July. Good for the West.

◄ **'Sungold'.** Origin: Minnesota. A selection from the same cross as 'Moongold', and the two must be planted together for pollination. The fruit is rounded and of medium size, with a tender, golden skin blushed orange. The flavor is mild and sweet, and the fruit is good fresh or preserved. The tree is upright, vigorous, and of medium size. The fruit ripens somewhat later than 'Moongold.' Good for all zones.

◄ **'Tilton'.** A vigorous tree that bears heavily most years. The fruit is yellow-orange and tolerates heat when ripening. It has a high chilling requirement (over 1,000 hours below 45°F).

◄ **'Wenatchee'.** The fruit is a large, flattened oval with orange-yellow skin and flesh. The tree does well in the Pacific Northwest. Good for the West.

Left: An espaliered cherry
Below: 'Black Tartarian'

CHERRIES

Cherries come in three distinct forms, with many varieties in each. The *sweet cherry* sold in markets is planted commercially in the coastal valleys of California and in the Northwest, especially Oregon. There are also extensive commercial plantings near the Great Lakes. All cherries require considerable winter chilling, which rules out planting in the mildest coastal and Gulf climates; but they are also damaged by early intense cold in fall, and by heavy rainfall during ripening. Sweet cherries are especially tricky for the home gardener, but try them wherever summer heat and winter cold are not too intense. *Sour cherries*, or *pie cherries*, are more widely adaptable and are good for cooking and canning. These are the most reliable for home gardeners, and there are many varieties developed for special conditions. The dwarf 'Meteor' and 'Northstar' pie cherries were developed for Minnesota winters. These and 'Early Richmond' and 'Montmorency' can all withstand both cold and poor spring weather better than sweet cherries. *Duke cherries* are hybrids with the shape and color of sweet cherries and the flavor and tartness of sour cherries.

Cherries come in many sizes. Bush varieties reach 6 to 8 feet tall and spread about as wide. The dwarf sour cherries grow to about 8 feet, but have a single trunk. Standard sour cherries and sweet cherries on dwarfing roots both reach 15 to 20 feet. A standard sweet cherry is the largest and can equal a small oak in size, if the climate permits. Such cherries can serve as major shade trees.

All cherries bear on long-lived spurs. Those on tree cherries can produce for 10 years and more, and begin to bear along 2-year-old branches. Count on the first crops in the third or fourth year after planting. Bush cherries may bear sooner.

Sweet cherries need a pollinator, with the exception of 'Stella'. 'Windsor' and 'Black Tartarian' are good pollinators and bear well, but always plant at least two varieties, or use a graft on a single tree. Sour cherries are self-fertile, as are bush cherries.

Cherries need no thinning and little pruning after the first two seasons of growth. See page 39 for developing a vase shape and apply this to young sour cherries. Sweet cherries may need heading back in the first years of growth to encourage branching.

Dwarf pie cherries have lovely flowers and make fine hedges and screens. They produce good crops and larger cherries can be grafted onto them for a choice of fruit and good pollination.

Birds are the major pests, but cherries also need protection from fruit flies, pear slugs (actually an insect larva), and bacterial leaf spot. See pages 29–33.

Check pollination requirements carefully if you plant a sweet cherry. For any cherry, check the recommended climate area. If you try a cherry outside its zone, offer protection in fall and winter.

Early Season Varieties

◀ **'Black Tartarian'.** This medium-size black cherry is fairly firm when picked, but softens quickly. It is widely planted because it is one of the earliest cherries and an excellent pollinator. Trees are erect and vigorous. Use any sweet cherry as a pollinator. Good for all zones. Widely available.

◀ **'Sam'.** Origin: British Columbia. This medium to large, black-fruited cherry is firm, juicy, and of good quality. The fruit resists cracking, and the tree is very vigorous, bearing heavy crops. Use 'Bing', 'Lambert', or 'Van' as a pollinator. Good for the North and West. Widely available.

Midseason Varieties

◀ **'Bing'.** This variety is the standard for black sweet cherries. The fruit is deep mahogany red, firm, and very juicy. It is subject to cracking and doubling. The tree is spreading and produces heavy crops, but suffers from bacterial leaf spot attack in humid climates. It is not easy to grow although it is quite popular. Use 'Sam', 'Van', or 'Black Tartarian' (*not* 'Napoleon' or 'Lambert') as a pollinator. Good for the West. Widely available.

◀ **'Chinook'.** Origin: Washington State. Like 'Bing' this variety has large, heart-shaped fruit, with mahogany skin and deep red flesh. The tree is spreading, vigorous, and a good producer. It is slightly hardier than 'Bing'. Use 'Bing', 'Sam', or 'Van' as a pollinator. Good for the West.

◀ **'Corum'.** Origin: Oregon. This variety is the recommended pollinator for 'Royal Ann' in the Pacific Northwest. It is a yellow cherry with a blush and thick, sweet, firm flesh. It is moderately resistant to cracking

Below: 'Stella'

Left: 'Royal Ann'

Below: 'Lambert'

and is a good canning cherry. The tree is fairly vigorous. Use 'Royal Ann', 'Sam', or 'Van' as a pollinator. Good for the West. Locally available in the Pacific Northwest.

◄ **'Emperor Francis'.** This large, yellow, blushed cherry resembles 'Napoleon' but is redder and more resistant to cracking. The flesh is very firm. The tree is very productive and hardier than 'Napoleon'. Use 'Rainier', 'Hedelfingen', or 'Gold' (*not* 'Windsor' or 'Napoleon') as a pollinator. Good for the North.

◄ **'Garden Bing'.** This genetic dwarf plant remains only a few feet high in a container, but grows to perhaps 8 feet in the ground. It is self-pollinating, and bears dark red fruit like 'Bing'. Good for the West.

◄ **'Gold'.** Origin: Nebraska. This is a yellow cherry that is highly resistant to cracking. Tree and blossoms are especially hardy and very productive. The tree withstands 30°F. Fruit of the original strain is small, but some strains have larger fruit. Use any sweet cherry as a pollinator. Good for the North and Midwest.

◄ **'Governor Wood'.** Light yellow with a red cheek, this cherry is large and rounded with tender, fine-textured flesh. Use 'Black Tartarian' or a pie cherry as a pollinator. Good for the South. Locally available.

◄ **'Kansas Sweet' ('Hansen Sweet').** Origin: Kansas. This is not really a sweet cherry but a fairly sweet form of the pie cherry group. The fruit is red and has firm flesh that is palatable fresh as well as in pies. The tree and blossoms are hardy in Kansas. It is self-fertile. Good for the North.

◄ **'Rainier'.** Origin: Washington. In shape, this cherry resembles 'Bing', but is a very attractive blushed yellow, with firm, juicy

flesh. The tree is vigorous, productive, and spreading to upright spreading. It is particularly hardy. Use 'Bing', 'Sam', or 'Van' as a pollinator. Good for the South and West.

◄ **'Royal Ann' ('Napoleon').** This very old French variety is the standard for yellow, blushed cherries. It is the major cherry used in commercial candies and maraschino cherries. The firm, juicy fruit is excellent fresh and good for canning. The tree is very large, extremely productive, and upright, spreading widely with age. The tree is relatively tender, although the buds tolerate some cold. Use 'Corum', 'Windsor', or 'Hedelfingen' as a pollinator (*not* 'Bing' or 'Lambert'). In hot summer areas double cherries are a severe problem. Good for all zones. Widely available.

◄ **'Schmidt'.** Origin: Germany. 'Schmidt' replaces 'Bing' as a major commercial black cherry in the East. The fruit is large and mahogany colored, with thick skin. The wine-red flesh is sweet but somewhat astringent. The large, vigorous tree is upright and spreading. It is hardy, but the fruit buds are fairly tender. Use 'Bing', 'Lambert', or 'Napoleon' as a pollinator. Good for the North and South. Widely available.

◄ **'Stella'.** Origin: British Columbia. This is the first true sweet cherry that is self-fertile (requiring no pollinator). The fruit is large, dark in color, and moderately firm. The tree is vigorous and fairly hardy and bears early. It can be used as a pollinator for any other sweet cherry. Good for the South and West.

◄ **'Van'.** Origin: British Columbia. This large, dark fruit has some resistance to cracking. The tree is very hardy and especially good in borderline areas, since it has a strong tendency to overset, and there-

fore may produce a crop when other cherries fail. It bears from 1 to 3 years earlier than 'Bing'. Use 'Bing', 'Lambert', or 'Napoleon' as a pollinator. Good for all zones. Widely available.

◄ **'Yellow Glass'.** This is an especially hardy variety with clear yellow fruit the size of a pie cherry, but sweet. Use 'Black Tartarian' as a pollinator. Good for the North.

Late Varieties

◄ **'Black Republican' ('Black Oregon').** This cherry is firm and very dark with slightly astringent flesh. The tree is quite hardy but tends to overbear heavily, producing small fruit. In borderline areas it may produce where others fail. Use any sweet cherry as a pollinator.

◄ **'Hedelfingen'.** Origin: Germany. The variety bears dark, medium-size fruit with meaty, firm flesh. One strain resists cracking, but some trees sold under this name do not. The tree is only moderately hardy, has a spreading and drooping form, and bears heavily. As a pollinator, use any sweet cherry listed here. Good for the North and South.

◄ **'Lambert'.** This large, dark cherry is similar to 'Bing' but ripens later. The tree is more widely adapted than 'Bing' but bears erratically in many eastern areas and is more difficult to train and prune. The strongly upright growth produces weak crotches if left untrained. Use 'Van' or 'Rainier' as a pollinator (*not* 'Bing', 'Napoleon', or 'Emperor Francis'). Good for all zones. Widely available.

Right: 'Valencia' orange
Below: 'Montmorency'

◀ **'Windsor'.** This is the standard late, dark commercial cherry in the East. The fruit is fairly small and not as firm as 'Bing' or 'Lambert'. However, its buds are very hardy, and it can be counted on to bear a heavy crop. A fine choice for difficult borderline areas where others may fail, the tree is medium size and vigorous with a good spread. For a pollinator, use any sweet cherry except 'Van' and 'Emperor Francis'. Good for the North and South. Widely available.

Duke Cherries
Duke cherries are hybrids of sweet and pie (or sour) cherries. They have tart, acid fruit like pie cherries, but are large, fairly upright trees like sweet cherries.

◀ **'Late Duke'.** This large, light red cherry ripens in late July. Use it for cooking or preserves. In cold climates it requires a sour cherry pollinator. In mild climates it is self-fertile. Good for the West.

◀ **'May Duke'.** This variety produces medium-size, dark red fruit of excellent flavor for cooking or preserves. In cold climates, use an early sweet cherry for pollination. In mild climates it is self-fertile. Good for the West.

Sour Cherries (pie cherries)
The following varieties are all self-fertile, and they will pollinate sweet cherries in mild climate areas. There are two types: the amarelle, with clear juice and yellow flesh; and the morello, with red juice and flesh. In the coldest northern climates, the amarelle is the commercial cherry.

◀ **'Early Richmond'.** An amarelle, the fruit is small, round, red, and excellent for pie, jam, and preserves. It is astringent when eaten fresh. The tree reaches 15 to 20 feet tall. Good for all zones. Widely available.

◀ **'English Morello'.** This late-ripening morello cherry is medium size, dark red, and crack resistant. The tart, firm flesh is good for cooking and canning. The tree has drooping branches and is small and hardy but only moderately vigorous and productive. Good for the North.

◀ **'Meteor'.** Origin: Minnesota. This amarelle is a genetic dwarf that reaches only about 10 feet tall. The fruit is bright red and large for a pie cherry, with clear yellow flesh. The tree is especially hardy but also does well in milder climates and is an ideal home garden tree for all cherry climates. Good for all zones. Widely available.

◀ **'Montmorency'.** This amarelle is the standard sour cherry for commercial and home planting. The large, brilliant red fruit has firm yellow flesh and is strongly crack resistant. The tree is medium to large, vigorous, and spreading. Various strains have slightly different ripening times and fruit characteristics. Good for all zones. Widely available.

◀ **'Northstar'.** Origin: Minnesota. This is a genetic dwarf morello, excellent for the home garden. It has red fruit and flesh and resists cracking. The tree is small, attractive, vigorous, and hardy and resists brown rot. Fruit ripens early but will hang on the tree for up to 2 weeks. Good for all zones. Widely available.

CITRUS
In much of the West and South there's nothing unusual about an orange tree outside the front door, bearing its juicy load of golden fruit. But even where winter cold makes it impossible to put citrus in the ground, a smart gardener can manage some of the pleasure of the perfumed flowers and bright fruits by planting in containers.

Citrus varieties can be grouped in two ways in relation to climate. First, for the gardener whose winter temperatures drop to 15° to 20°F the question is: Is the plant hardy enough? Of course you can always grow dwarfed varieties in containers and haul them off to shelter for the winter. The second question is: Will it get hot enough long enough for the fruit to develop sugar? Inadequate summer heat means sour fruit, so even in areas where the plants grow outdoors, a cool summer may make your mandarins very puckery. However, that doesn't matter at all for sour fruit like lemons, Rangpur limes, and so forth, and a sour mandarin is just as pretty on the tree and produces fine juice for drinks. The most climatically limited fruit is the grapefruit. If you want to eat grapefruit from the garden, you need desert heat.

Never avoid a plant because of climate. Citrus is a pleasure to look at and to smell, whether your crop is of market quality or not. We know a gardener in snow country who has a fine tubbed Mexican lime. Every winter she rolls it into her bright, south-facing kitchen and admires it more than if it grew outdoors.

All citrus are touchy about their roots and lower trunk. The plants need a soil with constantly available moisture, but

Right, top to bottom:
'Eureka' lemon, 'Meyer' lemon,
'Ponderosa' lemon
Below: *'Ruby' grapefruit*

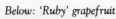

Below: 'Marsh' grapefruit

wet soil and poor drainage will be fatal in a short time. Warm-season gardeners should set their young plants high, as discussed on page 25, for bare-root trees. A thick mulch over the roots holds moisture and keeps temperatures down. If the subsoil is especially dense, you should raise the plants even more with a raised bed. Container plants need the very light mix mentioned above, so that you can water frequently without drowning the plant. Don't forget to feed heavily watered plants at least monthly with a food containing iron and zinc.

Whether in containers or in the ground, never bury the lower trunk of the plant. It should dry out immediately after watering, and if the plant is trimmed so the trunk is exposed to the sun, paint it with white interior latex paint to prevent sunburn.

In watering citrus, it is especially important not to wet the trunk and crown. Plant high and keep your watering basin at least a foot away from the crown.

Standard trees can reach 30 feet tall and wide. Dwarfed trees are a lot more manageable. A dwarfed navel orange grows to perhaps 10 feet, a mandarin to about 6 feet, and a kumquat to about 4 feet. Trees flower and fruit in the leaf nodes along newer growth, so the tree is decorated over the entire surface with bright orange, yellow, or yellow-green fruit. Since fruit will hang for months on many kinds of citrus, the fruit and bloom often occur together.

Citrus requires no attention to pollination. All kinds are self-fertile, and some need no pollination at all, forming seedless fruit without it.

Don't worry about thinning fruit unless you think a branch is about to break. Pruning should consist of removing any damaged branches (cut off winter-killed branches only after new growth is sturdy), and clipping off dead twiglets inside the tree. Sometimes an extra vigorous shoot suddenly sprouts beyond all the other branches. Reach in and break it off at the base. This will remove other buds that may do the same.

Dwarf citrus may sprout suckers from the rootstock, often a plant called trifoliate orange. Break these at the base as soon as you see them.

Pests and Disease
Stringent laws against transporting citrus from one commercial growing area to another help keep disease at a minimum. The worst thing a gardener has to face is root and crown rot caused by poor planting, poor soil, and standing water. Proper planting in light soil helps avoid this.

Check for hardiness, and prepare to protect tender plants by moving them in winter or covering them. Otherwise, experiment with any plant listed here.

Grapefruits
Grapefruit does best in the desert, although it ripens fairly well in any high-heat region. In cool areas it is thick skinned and pithy, but decorative.

◄ **'Marsh'.** A seedless, white-fleshed variety that takes 18 months to ripen, finishing in winter and spring in the desert. This beautiful large plant is not for cool, coastal climates.

◄ **'Ruby'.** A pink-fleshed fruit. Needs desert heat to color well.

Lemons
Lemons can't stand excessive heat or cold. Grow them in southern coastal areas or mildest northern areas with frost protection.

◄ **'Eureka'.** The commercial fruit. It ripens the whole year. The new growth is an interesting purplish color. Somewhat thorny.

◄ **'Lisbon'.** Takes more heat and cold than 'Eureka', and is good in the desert. Most fruit in fall, but some all year. Thorny.

◄ **'Meyer'.** A roundish yellow fruit with a rather sweet, aromatic peel and flesh. It is not quite a lemon. It tends to bloom and fruit constantly and is much hardier than lemon, good for coastal gardeners. The plant is small even on standard roots.

◄ **'Ponderosa'.** A conversation piece with enormous fruit. The tree is small and bears very young. Fruit ripens in winter.

*Center, top to bottom:
'Clementine' mandarin,
'Dancy' mandarin,
'Kinnow' mandarin,
'Valencia' orange*

Below: 'Mexican' lime

Below: 'Tarocco' orange

Limes

The lime is the most tender of all citrus. All limes need the mildest winters and high heat.

◄ **'Bearss'.** The hardiest: It grows wherever oranges grow. The fruit is lemon size, yellow when ripe, and ripens mainly in winter. Tree is thorny. Derived from a South Pacific variety.

◄ **'Mexican'.** This is the lime that appears in markets and is used in mixed drinks. The fruit is small, round, and green. The small tree is upright in habit. It will not take cold.

Mandarins

The mandarin is the small, loose-skinned fruit sometimes called tangerine or satsuma. The plants are very hardy, but the fruit needs high heat or it will be sour.

◄ **'Clementine'.** This needs a pollinator like 'Valencia', 'Dancy', or 'Orlando' tangelo. The fruit is fairly large with few seeds; ripens in late fall then hangs on for months. Needs less heat than other types.

◄ **'Dancy'.** The standard commercial fruit. Rather seedy but good for desert regions. Ripens in winter.

◄ **'Kara'.** Large fruit is slightly tart, ripens in winter in the desert, in spring elsewhere. Tree is spreading and drooping.

◄ **'Kinnow'.** Medium-size fruit with rich flavor, ripens from winter into spring. Tree is columnar. Good in any mild climate.

◄ **'Owari'. (Owari Satsuma).** Extremely hardy plant. The fruit is loose-skinned with delicate flavor. Ripens from October until winter. Trees are small with a spreading growth habit. Not for the desert.

Oranges

The orange tree is extremely attractive when it flowers and also when it fruits. The blossoms have a delicious fragrance and are often used in flower bouquets. The fruits vary a great deal and are put to a variety of uses.

◄ **'Shamoudi'.** This import from Israel resembles the navel orange, but grows on wide-spreading decorative trees.

◄ **'Tarocco'.** The flesh of this blood orange ranges from scarlet to orange with scarlet spots. Develops sugar without too much heat and is redder in the cooler growing areas. A very open tree.

◄ **'Valencia'.** This standard commercial juice orange ripens in summer and needs heat. Trees are full and vigorous. The fruit is thin-skinned and hard to peel, but the flesh is juicier than navel oranges.

◄ **'Washington'.** This is the standard navel fruit, available in markets from December to February. The fruit has relatively thick skin that peels easily, and a folded protrusion (the "navel") at the blossom end. A poor choice for desert regions, but it will ripen with less heat than other oranges. Trees reach 25 feet tall.

◄ **Sour orange.** This is used as a rootstock for many other citrus fruits. It lends itself to pruning as a hedge or screen, or you can grow it as a 30-foot tree. The small fruits are bitter, but many people like them in marmalade or in mixed drinks for a touch of bitter flavor.

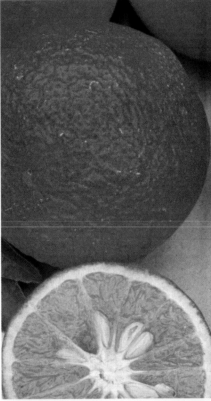

Tangelos

This is a mandarin—grapefruit cross with loose skin and the sugar of a mandarin but not the same aroma or flavor. Tangelos require cross-pollination from other citrus like the 'Valencia' orange.

◀ **'Minneola'.** A cross of 'Dancy' tangerine and 'Duncan' grapefruit. A large fruit with a prominent neck and tender flesh with much juice. Ripens in late winter. Needs desert or inland valley heat.

◀ **'Orlando'.** Same cross as 'Minneola'. Tender and juicy, ripening early, but needs heat for good flowering. It needs a pollinator like 'Dancy' or 'Clementine' mandarin.

Tangors

These are hybrids of tangerine (mandarin) and orange that need very high heat to ripen fruit properly.

◀ **'Temple'.** The best known commercial fruit. Good in the low desert and Florida. The orange-size fruit ripens in spring and is deep orange in color. Spreading, thorny tree.

Citrus Relatives

◀ **Calamondin.** Good as a container dwarf. It covers itself with tiny mandarin-like fruits that have very sour juice and a sweet rind. The plant is among the hardiest of citrus and has a decorative columnar shape. You can use the fruit for juice or marmalade. It is excellent for areas that are not good for citrus since you can protect it indoors during the worst cold, but won't lose it to early, mild frost.

◀ **Kumquat.** Not a citrus, but a member of the related genus *Fortunella*. It has tiny round-to-oval fruit with sour flesh and sweet rind. The foliage is attractive and the tree is small even on standard roots. Use it for preserves or eat it fresh, skin and all.

◀ **Pomelo.** A grapefruit relative that has large fruits, much bigger than any other. You eat it by peeling both the rind and the skin of each segment. May be hard to find in nurseries.

◀ **'Rangpur' Lime.** Really a very sour mandarin on an attractive plant that is easy to espalier. Use it for juice, but don't expect lime flavor. It bears heavy crops and is very hardy; good in coastal areas.

CRAB APPLES

Fine for jellies or pickled whole fruit, crab apples are also the most decorative of fruit trees. Flowers range from red to pink and white. Leaf color may be red, bronze, variegated red and green, or green. Fruits are of many sizes, from tiny cherrylike varieties to large, yellow, pink-cheeked kinds. The varieties sold for their flowers also have edible fruit, but the large-fruited varieties are better if your aim is to grow the fruit for jelly.

Crab apples range from small, 10-foot trees to spreading trees 25 feet tall. The large-fruited kinds are larger trees.

Crab apples fruit on long-lived spurs, generally producing clusters of several fruits on each spur. Since crops are heavy, you can cut back new wood without losing anything.

Crab apples are self-fertile, but you can graft several kinds that bloom at different times to extend the flowering season.

Train the young trees to a vase with 3 or 4 scaffolds. After the second year you can leave them alone or cut them back to maintain size. No thinning of the crop is necessary.

Use small varieties if you're crowded. If you have no space but want a light crop for jelly, graft a branch to an existing apple tree.

Crab apples are subject to the same diseases as apples, and scab is a major problem for some varieties. Choose resistant kinds.

Varieties

We include both large-fruiting kinds and those that are mainly ornamental, but also offer a good crop of smaller fruits. Use red or pink fruits if you want pink jelly.

◀ **'Barbara Ann'.** This ornamental offers dark, reddish-purple, ½-inch fruit with reddish pulp. The tree produces a profusion of 2-inch, purple-pink, full, double flowers. It grows to about 25 feet tall and is reasonably disease resistant.

◀ **'Chestnut'.** This very large, bronzy-red crab apple is big enough to make a good dessert or lunchbox fruit, and can also be used to make a deep pink jelly. Its flavor is especially pleasing. The tree is very hardy, medium size, and reasonably disease resistant.

◀ **'Dolgo'.** Origin: Russia. The smallish, oblong red fruit is juicy and if picked before fully ripened it jells easily, making a ruby red jelly. The tree is hardy, vigorous, and productive. Fruit ripens in September. Widely available.

◀ **'Florence'.** The large yellow fruit has an attractive red blush. Use it for pale pink jelly or for pickling whole. The tree is medium size and somewhat tender, so it is best planted in warmer regions. It ranges from fairly to very productive. Widely available.

◀ **'Hyslop'.** This medium-size fruit is yellow blushed with red. Use it for whole relishes or for pale pink jelly. The tree is fairly hardy and ornamental, with single pink flowers.

◀ **'Katherine'.** The tiny fruits of this variety are an attractive yellow with a heavy red blush. They can be made into a pink jelly. The tree is small, slow growing, and fairly hardy, but it flowers and fruits only every other year. It grows about 15 feet tall and is reasonably disease resistant. It has double flowers that open pink and then fade to white.

◀ **'Montreal Beauty'.** This medium-size green crab apple with red striping makes good jelly on its own or is a good base for mint or rose geranium jellies. The tree is medium to large, hardy, and fairly disease resistant. Locally available.

◀ **'Profusion'.** The tiny scarlet fruits of this variety are good in jellies. The tree spreads only slightly, is small (about 15 feet), and produces small single flowers that are deep red in bud and open to purplish red to blue pink. It is moderately susceptible to mildew.

◀ **'Siberian Crab'.** This variety bears an abundance of clear scarlet, medium-size fruit that can be jellied or pickled whole. The tree is vase-shaped and reaches 15 to 30 feet tall, depending on climate and soil. The inch-wide white flowers are fragrant. Some strains are disease resistant and some are not.

◀ **'Transcendent'.** These large yellow crab apples are blushed with pink on one side. Use them for clear jellies or eat them fresh if you like the wild, astringent flavor. The tree is medium to large and somewhat disease resistant, but not very hardy, so plant only in warm areas.

◀ **'Whitney'.** An old favorite, this variety has very large fruit, good for fresh eating, jelly, preserves, and apple butter. The fruit is yellow with red stripes. The tree is hardy, medium to large, and reasonably disease resistant. Widely available.

◀ **'Young America'.** The large and abundant red fruit on this variety makes a clear red jelly with a splendid flavor. The tree is especially vigorous and hardy, and the fruit ripens about mid-September.

Left: A heavily pruned old fig tree

Below: 'Brown Turkey'

Below: 'Mission'

FIGS

Although the fig is generally thought of as a subtropical fruit suited mainly to the mild winters and heat of the Gulf Coast regions, California, and the Southwestern desert areas, some varieties will bear even in the milder climates of the Northwest and Northeast. If a freeze knocks the plant down, it will sprout again quickly.

Figs are good fruit for gardeners who don't want to fuss with spraying and pruning, but like to see a big crop on their tree. In warm regions the fig bears big, juicy fruit in early summer, then sets a heavier crop of small fruit, perfect for drying, in the fall. It lives for many years, loves clay soil if drainage is good, and needs next to no attention. You have a choice of dark fruit with red flesh or greenish-yellow fruit with bright pink flesh.

In cold-winter regions, fig shrubs reach 10 feet tall and spread that much or more. In warm regions, trees reach 15 to 30 feet and spread wide and low, but you can easily cut them back or confine them.

Figs are not really fruits in the botanical sense. They are a collection of closed-up flowers with all the important parts accessible to the outside world only through a hole at the base. The first crop "blooms" on new wood of the previous season, and the second crop appears on new wood of the current season. When a tree is cut back to confine it, you usually lose most of the first crop. You also lose it in cold regions where winter does your pruning.

Most home garden varieties need no pollination: The California commercial fig, 'Calimyrna', does need pollinating and should not be planted in the South.

No fruit thinning is necessary. Prune the young tree to an open shape in the first two years, then remove any suckers at the base when you think about it. Pull —don't cut—them. Shrub forms need no attention except for removing dead wood.

You can prune a fig to a 5-foot spreading shrub if you like, as discussed on page 42, or flatten it on a south wall. Figs can also be grown as container plants for use on a patio, allowing them to be protected in winter by moving them to a garage or storage area.

Figs need no attention to pests or disease.

Varieties

◄ **'Adriatic'.** The fruit is green skinned with a strawberry-colored pulp. In hot areas, the second crop has a paler pulp, and in cool areas the fruit of both crops is larger. This is principally a drying fig. The tree is vigorous and large. Used in California for drying and processing into figbars. Locally available.

◄ **'Brown Turkey' ('Turkey', 'Southeastern Brown Turkey', 'San Piero', 'Black Spanish').** A large-fruited variety for fresh use. The first crop is large and dark brown, the second crop is smaller. The pulp is light strawberry. The tree is small and can be pruned heavily to cut the crop. One of the hardiest varieties, it is worth trying in the Northeast and Northwest. Recommended for the Southeast and Southwest.

◄ **'Celeste' ('Blue Celeste', 'Celestial', 'Sugar', 'Malta').** Bronzy fruit with a violet tinge. The pulp is amber with rose tones. 'Celeste' is the most widely recommended fig in the Southeast, but is also grown in the West. An especially hardy plant.

◄ **'Conadria'.** Origin: California. One parent is 'Adriatic'. The fruit is thin skinned and white with a violet blush.

The red flesh resists spoilage. The tree is vigorous and precocious, producing two crops. Recommended for the hot valleys of California. Locally available.

◄ **'Kadota' ('Florentine').** Fruit is toughskinned and greenish-yellow, and the first crop has a richer flavor. This is principally a canning and drying variety. The tree is vigorous. Recommended for hot California valleys.

◄ **'King' ('Desert King').** The fruit is green with flecks of white; the pulp violet pink. The tree comes back from the roots after a freeze and bears in fall. Recommended for Oregon fig climates. Locally available.

◄ **'Latterula' ('White Italian Honey Fig').** This large greenish-yellow fig with honey-colored pulp grows on a very hardy tree that bears two crops. Recommended for Oregon fig climates. Locally available.

◄ **'Magnolia' ('Brunswick', 'Madonna').** This is a large straw-colored fig on a fairly hardy tree. Recommended for the Southeast.

◄ **'Mission' ('Black Mission').** This variety bears two heavy crops of black fruit with deep red pulp. The first crop has larger fruit; the second crop can be dried. The tree is large and vigorous. Recommended for California and desert regions, but also grown in warmer Southeastern zones.

◄ **'Texas Everbearing' ('Dwarf Everbearing').** The fruit and tree resemble 'Brown Turkey'. This variety will resprout and bear after a freeze kills back the top. Recommended for the Southeast and South Central.

Also check the local availability of 'Granata', 'Negronne', and 'Neveralla' in Oregon; 'Genoa' ('White Genoa') on the California coast; and 'Green Ischia' and 'Hunt' in the Southeast.

PEACHES

The peach is one of the most popular of homegrown fruits. Both peaches and their close relatives, the nectarines, are at their best when tree ripened, so a home gardener's time and effort are rewarded by a product that money can't buy.

Peaches cannot tolerate extreme winter cold or late frost, so in the northern plains states and northern New England, peaches are purely experimental. The hardiest, such as 'Reliance', may survive and bear in a protected spot, but you can't be sure. Peaches do well in more temperate climates near the Great Lakes, but choose the warmest site available for planting. A protected sunny spot where cold air can't collect and sit is the right place for your tree.

Some of the greatest peach-growing country in the world is in the West: California alone produces 50 percent of the commercial peaches in the United States. Peaches do well in the San Joaquin Valley in California, in South Carolina and semicoastal areas of the East, and in some areas of eastern Washington. To produce great peaches the climate must fulfill high-chill requirements (700 to 1,000 hours of cold winter weather at 45°F or below) unless otherwise stated. This should be followed by warm, dry spring weather and hot summers. Gardeners not blessed with this prime climate can grow satisfactory fruit by selecting the right varieties for their own gardens. Selected low chill varieties can fruit well in all subtropical climates but southern Florida.

The standard tree on a peach rootstock grows to about 18 feet tall and 15 feet wide. It could grow larger if left alone, but it is best pruned heavily each year to maintain that size and to encourage lots of new growth along the branches. A semidwarf tree on 'Nanking' cherry or 'St. Julien' plum rootstock stays 6 to 8 feet tall, but it is difficult to keep it healthy and productive. The genetic dwarf peaches grow in bush shape to about 4 to 7 feet tall and require no pruning to maintain size or to force growth. At most, you will clip out tangles and remove broken twigs. These are a good choice for patio container plantings.

You won't have any trouble fitting a peach to whatever space is available to you. Commercial growers of standard trees plant peaches in a hedgerow, a narrow wall of trees formed by training the young trees to a V-shaped scaffold, then pruning to hold all growth within the hedge shape. This means trimming a bit in summer as well as pruning in winter, but it's not difficult. Just clip any runaways that shoot beyond the general hedge outline to about 8 to 10 leaves long. See page 48 for a trellis version of the same V-shape.

In the home garden peach trees can be planted 2 or 3 to a hole if the varieties are well chosen. This will spread the harvest over several weeks. Dig an extra-big planting hole and set 2 to 3 varieties together, with their roots almost touching. Also, you can graft different limbs to different varieties and have 3 varieties on a single tree.

Fruit Growth

The sketch on page 41 shows you the most important thing about a peach tree: It illustrates where on the branch you'll find flowers and fruit. Notice that fruit is formed only on branch segments that grew the previous summer. New wood grows on beyond the fruit and will produce next year's crop. Unlike apricots and plums, once a peach is harvested, the section of branch on which it grew will never fruit again. Encourage new growth for replacement branches by pruning heavily every winter.

You encourage new growth on standard or semidwarf trees by pruning. As you prune, you thin, head back, and remove weak branches. The tree responds with lush growth. See page 42 for details and a sketch of cuts to make.

Once a crop sets on a peach tree, you may not even see the branches through all the fruit. You can't leave it all because it will be too small; it slows branch growth; and it may snap branches. Thin it out when it reaches thumbnail size. For early season peaches, leave 6 to 8 inches of space between each fruit; for late season peaches, leave 4 to 5 inches between each fruit. If a frost knocks off much of your crop, leave all the remaining fruits, even if they're clustered. What is important is the ratio of leaf surface to peaches, so a sparse crop will do equally well in singles or bunches.

Peaches are twiggy trees, but the greatest number of flower buds form on sturdy new branches that made more than a foot of new growth the previous summer. Keep these and thin the more anemic twigs. You can head the strong ones back by one-third to one-half if you want to keep the tree small. They'll bloom on the remaining half.

Only a few peach varieties need a pollinator. Normally the trees are self-fertile, although bees are a big help in pollen transfer.

Left: 'Redhaven'
Below: 'Veteran'

All peaches like a winter rest. Without it, they bloom late, open their leaves erratically, and finally die. Be sure to choose varieties that suit your climate. Southern California peaches are listed as such. Some peaches have been bred for short, mild winters and may bloom too early or may freeze in the North. Be sure to buy hardy, high-chill varieties for the North. Chill is an important factor in the South as well. A high-chill peach will leaf out and flower erratically in southern Mississippi, while a low-chill peach may try to bloom before the last frost in Tennessee.

The universal peach ailment is leaf curl, but you can control it easily with a copper spray. See pages 32–33 for control methods. You will also probably encounter the peach tree borer, gnawing the trunk at ground level. Brown rot attacks fruit but is controllable. For bacterial spot on leaves, check the list for resistant varieties. Brown rot and plum curculio are the chief pests in the North.

Very Early Varieties
◀ **'Desert Gold'.** Origin: California. This medium-size, round fruit has yellow skin with a red blush. The flesh is yellow, firm, and semifreestone. The tree is fairly vigorous and productive and requires heavy thinning. The chilling requirement is very low—200–300 hours. Good for the desert and coastal areas of the West.
◀ **'Springtime'.** Origin: California. The small- to medium-size fruit has yellow skin with a high blush and abundant short fuzz. The flesh is white and semifreestone. Good for the West.
◀ **'Tejon'.** Origin: California. The medium-size fruit is yellow with a red blush over half its surface; light fuzz. The

yellow flesh is semifreestone. The tree bears very well. Good for the West, particularly Southern California. Locally available.

Early Varieties
◀ **'Candor'.** Origin: North Carolina. An outstanding variety for its season, this is a medium-size semifreestone. The skin is bright red over a yellow ground. The flesh is yellow, fine textured, and nonbrowning. It is productive and good for canning. The foliage has moderate resistance to bacterial spot. Good for the South.
◀ **'Dixiered'.** Origin: Georgia. This medium-size cling peach usually escapes spring frost damage. The skin is bright red, and the flesh is yellow flecked with red, firm, fine textured, and nonbrowning. The tree is vigorous and consistently productive. Good for the South.
◀ **'Earlired'.** Origin: Maryland. This is a medium-size cling peach. The skin is yellow blushed with red, and the flesh is yellow, firm, and medium textured. Thin early and heavily. It is moderately susceptible to bacterial leaf spot. Good for the South.
◀ **'Erly-Red-Fre'.** Origin: Virginia. This large, semifreestone, red peach has white flesh. The tree is vigorous and hardy and production is average. This variety is resistant to bacterial leaf spot. Good for the South.
◀ **'Fairhaven'.** Origin: Michigan. This large peach is bright yellow with an attractive red cheek and light fuzz. The firm freestone flesh is yellow with red at the pit. The fruit freezes well. The tree has showy flowers. Good for the West.
◀ **'Flavorcrest'.** Origin: California. A large, firm, yellow freestone with good flavor. The skin is blushed red. Good for California.

◀ **'Garnet Beauty'.** Origin: Ontario, Canada. This variety is an early sport of 'Redhaven'. Medium to large, semifreestone fruit hangs on the tree until overripe. The firm flesh is yellow streaked with red and is slightly fibrous in texture. The tree is vigorous and hardy and produces heavy crops that achieve good size and color even inside the tree. It is susceptible to bacterial leaf spot. Good for the North. Widely available.
◀ **'Golden Jubilee'.** Origin: New Jersey. An old standby, this medium to large freestone has skin mottled bright red. The flesh is yellow, firm, and coarse in texture. The tree is hardy and sets heavy crops but is self-thinning. Good for all zones. Widely available.
◀ **'Redhaven'.** Origin: Michigan. One of the finest early peaches, this medium-size freestone is widely recommended. The skin is deep red over a yellow ground. The flesh is yellow, firm, and nonbrowning. Fruit sets heavily and is good for freezing. This tree needs heavy thinning but rewards with outstanding fruit. The tree is spreading, vigorous, and highly productive, and resists bacterial leaf spot. Good for all zones. Widely available.
Note: 'Early Redhaven' is nearly identical, but two weeks earlier.
◀ **'Redtop'.** Origin: California. The large fruit is nearly covered with an attractive blush and light fuzz. The yellow freestone flesh is unusually firm and good canned or frozen. The tree is moderately vigorous

Left: 'Loring'
Below: 'Early Elberta'

Below: 'Georgia Belle'

and somewhat susceptible to bacterial leaf spot. Flowers are showy. Good for the West.

◄ **'Reliance'.** Origin: New Hampshire. A promising home garden variety, this tree is very winter hardy: It will withstand −20° to −25°F in January and February and will still produce a crop that same year. The large freestone fruit has dark red skin over a yellow ground. The flesh is bright yellow, medium firm, and slightly stringy. Flowers are showy. Good for the North and West. Widely available.

◄ **'Springcrest'.** Origin: California. A medium-size, flavorful, yellow freestone. The tree is vigorous and productive and has showy flowers. The fruit matures in late May. Good for the West.

◄ **'Sunhaven'.** Origin: Michigan. A medium to large peach. The skin is bright red over a golden ground and has short, soft fuzz. The flesh is yellow flecked with red, firm, fine textured, and nonbrowning. The tree is vigorous and consistently productive. Recommended for all zones. Widely available.

◄ **'Veteran'.** Origin: Ontario, Canada. A favorite in western Washington and Oregon, the medium to large fruit is yellow splashed with red and has medium fuzz. The nearly freestone flesh is yellow and soft. The tree is vigorous and highly productive—one of the very best in cool Pacific climates. Good for the West.

◄ **'Ventura'.** Origin: California. A good low-chill (400 hours below 45°F), yellow-fleshed freestone for southern California. The tree has average vigor and productiveness. The peach has yellow skin with a red blush, good flavor, and firm flesh. Available in southern California.

Midseason Varieties

◄ **'Babcock'.** Origin: California. The name 'Babcock' is often applied to a range of early white peaches, especially in the market. The small- to medium-size fruit is light pink, blushed red with little fuzz. The skin peels easily. The nearly pure white flesh is red near the pit, tender, and juicy and has a mild flavor. The medium to large tree is spreading and vigorous, but needs heavy thinning early in the season to produce large fruit. Good for the West, particularly southern California.

◄ **'Early Elberta'** (Gleason Strain). Origin: Utah. This large freestone matures 3 to 10 days before 'Elberta'. The flesh is yellow and is considered of better flavor than 'Elberta'. It is good for canning and freezing. The tree is hardy and consistently productive. Good for the South and West. Widely available.

◄ **'Glohaven'.** Origin: Michigan. This large freestone peach has skin that is red over deep yellow. The flesh is yellow, firm, and nonbrowning. It cans and freezes well. Flowers are medium size and deep pink. The tree is vigorous and hardy. The fruit remains on the tree when mature. Good for the North. Widely available.

◄ **'J. H. Hale'.** The skin of this extra-large freestone is deep crimson over a yellow ground and nearly fuzzless. The flesh is golden yellow and firm. This variety needs cross-pollination for best production. Good for all zones. Widely available.

◄ **'July Elberta'** ('Kim Elberta'). Origin: California. This variety is best suited for the Willamette Valley. The medium-size fruit is greenish yellow, blushed and streaked with dull red, and very fuzzy. The yellow flesh is of high quality. The tree is vigorous and bears heavily but is susceptible to bacterial leaf spot. Good for the West.

◄ **'Loring'.** Origin: Missouri. The skin of this medium-size freestone has a slight fuzz and is blushed red over a yellow ground. The flesh is yellow, firm, and medium textured. It is a dependable cropper, setting fruit under adverse weather conditions, and resists bacterial leaf spot. Good for the North and South. Widely available.

◄ **'Redglobe'.** Origin: Maryland. Very adaptable to many peach-producing areas, this variety is above average quality for canning and freezing. A medium to large freestone, its skin is bright red over yellow ground, and the flesh is yellow, firm, and fine textured. It is a dependable producer. Showy flowers are deep pink. It is susceptible to bacterial leaf spot. Good for the South and West. Widely available.

◄ **'Suncrest'.** Origin: California. A firm, large freestone with a red blush over yellow skin. This variety is susceptible to bacterial leaf spot and should be grown in the West and areas without this disease. The vigorous tree often bears large crops. It is hardy in cold sections of the North. Widely available.

Late Varieties

◄ **'Blake'.** Origin: New Jersey. This large freestone has red, slightly fuzzy skin. The flesh is yellow and firm. The fruit hangs well on the tree, which is vigorous and productive. 'Blake' is good for freezing and excellent for canning. It is susceptible to bacterial canker. Good for the North and South.

◄ **'Cresthaven'.** Origin: Michigan. The skin of this medium to large freestone is bright red over a gold ground and almost fuzzless. The flesh is yellow and nonbrowning. The tree is hardy, and fruit stays on

Left: 'Madison' peach
Below: 'Earliblaze' nectarine

when mature. It is good for canning and freezing. Good for the North and South. Widely available.

◀ **'Elberta'.** This large freestone is the old favorite for a midseason crop. The skin is red blushed over a deep golden yellow ground color. The fruit tends to drop at maturity. The slight bitterness around the pit is preferred by many. It is resistant to brown rot. Good for all zones. Widely available.

◀ **'Georgia Belle'.** Origin: Georgia. This is an outstanding white peach. The skin is red blushed over creamy white. The flesh is white and firm, has excellent flavor, and is fair for freezing, but poor for canning. The tree is vigorous, very winter hardy, and productive, but very susceptible to brown rot. Good for the North and South. Widely available.

◀ **'Jefferson'.** Origin: Virginia. Especially suited to localities where late spring frosts are a problem, this peach is noted for its fine texture and flavor. The skin is bright red over a bright orange ground. The flesh is yellow and firm. It is a reliable producer that cans and freezes well. It has some resistance to brown rot. Good for the North and South.

◀ **'Jerseyqueen'.** Origin: New Jersey. This large freestone peach is bright red over a yellow ground. The flesh is yellow and firm. Flowers are showy, and the tree is productive. This tree is winter hardy in the North, but it is susceptible to bacterial leaf spot. Good for the South. Widely available.

◀ **'Madison'.** Origin: Virginia. Adapted to the mountain areas of Virginia, this variety has exceptional tolerance to frosts during the blossoming season, setting

crops where others fail. The skin of this medium-size freestone fruit is bright red over a bright orange-yellow ground. The flesh is orange-yellow, very firm, and fine in texture. The growth of the tree is average to vigorous. Good for the North and South. Widely available.

◀ **'Raritan Rose'.** Origin: New Jersey. This vigorous, winter-hardy tree produces delicious white-fleshed, freestone peaches. The skin is red. Available in the East and North.

◀ **'Redskin'.** Origin: Maryland. A popular peach that ripens after 'Elberta'. It has good red color, is firm, and handles well. The flesh is nonbrowning, making this an excellent peach for freezing, canning, and eating fresh. Widely available in the East and North.

◀ **'Rio Oso Gem'.** Origin: California. The skin of this large freestone is red over a yellow ground; the flesh is yellow, firm, fine in texture, and nonbrowning. It is good both fresh and for freezing. Blossoms are light pink, very large, showy, and appear later than most peach blossoms. The tree is productive but not vigorous. Good for the South and West. Widely available.

◀ **'Sunhigh'.** Origin: New Jersey. This is a very good medium to large freestone. The skin is bright red over a yellow ground. The flesh is yellow and firm. The tree is vigorous and spreading. It is very susceptible to bacterial leaf spot and requires thorough summer spraying. Good for the North and South. Widely available.

NECTARINES

The nectarine is simply a fuzzless peach. Peach trees sometimes produce nectarines as sports, and nectarine trees will produce fuzzy peach sports. The two plants are nearly identical, but nectarines are generally more susceptible to brown rot. Gardeners in the South may have trouble with the disease because hot humid weather encourages it. You will have to spray regularly to control it. Otherwise, nectarines require the same care as peaches.

Early Varieties

◀ **'Earliblaze'.** Origin: California. This is a medium-size, clingstone, yellow-fleshed fruit that ripens ahead of the 'Redhaven' peach. It has red skin and a prominent suture (fold down the length of the fruit). Good for the North and South.

◀ **'Independence'.** Origin: California. This medium-size, oval fruit has brilliant cherry red skin. The freestone flesh is yellow and firm. The tree is productive and moderately vigorous, with showy flowers. It will take warm winters. Good for the South and West.

◀ **'Pocahontas'.** Origin: Virginia. The medium to large, oval fruit is a bright red. The semifreestone flesh is yellow, slightly stringy, and of good quality. This variety resists brown rot and frost during the blossoming season. The flowers are not showy. Good for the North and South.

◀ **'Silver Lode'.** Origin: California. The skin of this fruit is red. The freestone flesh is white and sweet and of good texture. The tree requires little chilling. Good for the South and West.

'Redgold'
Below: 'Fantasia'

◄ **'Sungold'.** Origin: Florida. This medium-size freestone has red skin and firm yellow flesh. It is a moderate-chill (555 hours below 45°F) variety and has some resistance to brown rot. Good for the South.

◄ **'Sunred'.** Origin: Florida. This low-chill nectarine (300 hours below 45°F) is adapted to Florida and ripens there in May. It is a small, yellow-fleshed cling-stone with red skin.

Midseason Varieties

◄ **'Fantasia'.** Origin: California. This fairly large fruit has bright yellow skin covered up to two-thirds with a red blush. The freestone flesh is yellow, firm, and smooth. The tree is vigorous and produc-tive, with showy flowers. It requires moderate-chill (500—600 hours below 45°F) and is susceptible to brown rot and bacterial leaf spot. Good for the South and West.

◄ **'Flavortop'.** Origin: California. The large oval fruit is mostly red, with firm, smooth, freestone, yellow flesh. The tree is vigorous and productive with showy flowers, and needs moderate winter cold. It is susceptible to brown rot and bacterial leaf spot. Good for the South and West.

◄ **'Garden State'.** Origin: New Jersey. The large fruit has a rounded oval shape. The skin is greenish-yellow, almost completely covered with red. The yellow freestone flesh is firm and juicy. The tree is vigorous and spreading, with large showy flowers. Good for all zones.

◄ **'Mericrest'.** Origin: New Hampshire. An extremely winter hardy, yellow-fleshed, red-skinned nectarine with a large suture. It resists bacterial leaf spot and brown rot. Good for the North.

◄ **'Nectared 4'.** Origin: New Jersey. The fairly large fruit is yellow with a red blush over much of the surface. The semifree-stone flesh is yellow. The tree is produc-tive, with showy flowers. Good for the North and South.

◄ **'Nectared 5'.** Origin: New Jersey. The large fruit is smooth, with a blush cover-ing most of the yellow skin. The yellow flesh is semifreestone until fully ripe, then freestone. The tree is productive. Good for the South.

◄ **'Panamint'.** Origin: California. The fruit has a red skin, and the freestone flesh is yellow. The tree is vigorous and productive and needs little winter chill-ing. Good for the South and West.

◄ **'Pioneer'.** Origin: California. The fruit has a thin red skin. The freestone yellow flesh is red near the pit and has a rich, distinctive flavor. The tree requires little chilling and has large, showy blossoms. Good for the West. Locally available.

◄ **'Redchief'.** Origin: Virginia. This medium fruit is bright red and attractive. The flesh is freestone, white, and fairly firm. The tree is vigorous and productive, with showy flowers, and is very resistant to brown rot. Good for the South.

◄ **'Redgold'.** Origin: California. A hardy, firm, freestone nectarine with glossy red skin. It resists brown rot and cracking but is susceptible to mildew. Good for the North and South.

◄ **'Sunglo'.** Origin: California. This heavy-bearing variety requires heavy thin-ning and extra nitrogen for best perfor-mance. This medium-size freestone has bright red over golden orange skin with firm flesh. Good for the North.

Late Varieties

◄ **'Cavalier'.** Origin: Virginia. The medium fruit is orange-yellow with splashes and mottles of red. The yellow freestone flesh is firm, aromatic, and slightly bitter. The vigorous and produc-tive tree has showy flowers and resists brown rot. Good for the North and South.

◄ **'Fairlane'.** Origin: California. A very late ripening, red-skinned, yellow cling-stone. Good for the West.

◄ **'Flamekist'.** Origin: California. A large, red-skinned, yellow-fleshed, clingstone nectarine. It has moderate-chill require-ments (500—600 hours below 45°F). It is susceptible to brown rot and bacterial leaf spot. Good for the West.

◄ **'Gold Mine'.** An old variety. The large white fruit is blushed red. The juicy white freestone flesh has an excellent flavor. It is a moderate-chill variety (500 hours below 45°F). Good for the West.

◄ **'Late Le Grand'.** Origin: California. The large clingstone fruit has yellow skin with a light red blush. This was the first large, firm, yellow commercial nectarine. The spreading tree is productive, with large, showy flowers. It is susceptible to brown rot and bacterial leaf spot. Good for the West.

*Left: 'Nectarina'
dwarf nectarine*

*Below: 'Bonanza'
dwarf peach*

GENETIC DWARF PEACHES AND NECTARINES

The genetic dwarf peaches and nectarines form dense bushes, with long leaves trailing in tiers from the branches. In spring the branches are entirely hidden by flowers that are usually semidouble and always very showy. In winter the bare plants are also visually interesting. The plants are not winter hardy and must be moved indoors for the season in northern climates. The fruit of all these plants is normal size.

In containers, the plants can be kept to about 5 feet tall, but in the ground they will eventually reach 6 to 8 feet and spread 6 to 9 feet. They can be used as ornamentals and require minimal pruning. The fruit flavor and texture are not as good as standard-size trees so they are not used commercially. The fruit must be thinned, and the trees need normal spraying for all the peach diseases and pests. These dwarfs are more susceptible to mites than normal-size peach trees.

Most require moderate winter chilling (about 400–600 hours below 45°F) for good bloom, and none are blossom hardy in really cold places, but they can be grown in containers and protected until the warm season. If you try this method in the coldest northern regions, you may have to pollinate the flowers yourself with the eraser on a pencil, touching it first to pollen, then to the stigma of a different flower.

These dwarf plants were created by breeding numerous varieties, but they all probably share the common heritage of the 'Swataw' peach or the 'Flory' peach, Chinese genetic dwarf varieties.

Genetic Dwarf Peaches

◀ **'Bonanza'.** Origin: California. A medium-size, yellow-fleshed freestone with red blush, this was the original genetic dwarf peach developed for the home gardener from earlier dwarfs like 'Flory'. It has a moderate chilling requirement (about 500 hours below 45°F), and the fruit ripens in mid-June in California. Good for the West and South.

◀ **'Compact Redhaven'.** Origin: Washington. This tree is larger than other dwarfs (up to 10 feet), and its leaves and growth habit resemble standard trees more than genetic dwarfs. The fruit resemble 'Redhaven' in size, quality, and color but are borne on a more compact tree. It tolerates cold better than genetic dwarfs. Good for all zones, especially the North, Midwest, and East.

◀ **'Empress'.** Origin: California. A medium-size, yellow-fleshed clingstone with glowing pink skin. It has sweet flavor, juicy texture, and a moderate chilling requirement (500–600 hours below 45°F) and ripens in early August in California. Good for the West and South.

◀ **'Garden Gold'.** Origin: California. A large, yellow-fleshed freestone with red skin and cavity. A moderate-chill variety with showy flowers, the fruit ripens in mid-August in California. Good for the West and South.

◀ **'Garden Sun'.** Origin: California. A large, yellow-fleshed freestone with red skin and cavity. A moderate-chill variety, the fruit ripens in early August in California. Good for the West and South.

◀ **'Honey Babe'.** Origin: California. A large, firm, orange-fleshed freestone with red skin. The fruit has distinctive peach flavor and rates high for flavor and sweetness. It is a moderate-chill variety, ripening before 'Redhaven', mid-June in California. Good for the West and South and worth trying in the East with protection.

◀ **'Southern Flame'.** Origin: California. A large, yellow freestone with red skin and cavity. A good eating quality fruit that ripens in late July in California. Low chilling requirement (about 400 hours). Good for the West and South.

◀ **'Southern Rose'.** Origin: California. A large, firm, yellow-fleshed freestone with red blush. Rated as low-chill (300–400 hours below 45°F), the fruit ripens in early August in California. Good for low-chill areas of the West and South.

◀ **'Southern Sweet'.** Origin: California. A medium-size, yellow-flesh freestone with red blush and good flavor. This moderate-chill variety matures in mid-June in California, ahead of 'Redhaven'.

◀ **'Sunburst'.** Origin: California. A large, firm, yellow-fleshed, clingstone with red blush. The fruit is juicy with a red cavity, has good flavor, and ripens in mid-July. It is a high-chill variety suggested for warm areas of the East and South and colder areas of the West.

Genetic Dwarf Nectarines

◀ **'Garden Beauty'.** Origin: California. This yellow-fleshed clingstone with red skin has a low chilling requirement (about 400 hours below 45°F). It has large, double flowers and ripens in late August in California. Good for the South and West.

◀ **'Garden Delight'.** Origin: California. This yellow-fleshed freestone has red skin. The fruit ripens in mid-August in California. Good for the South and West.

◀ **'Garden King'.** Origin: California. This yellow-fleshed clingstone with red skin has a low chilling requirement and ripens in mid-August in California. Good for the South and West.

◀ **'Golden Prolific'.** Origin: California. This large, yellow-fleshed freestone with orange skin and a red center has a high chilling requirement (900 hours below 45°F). The fruit ripens in late August. Good only for high-chill areas in the West but worth trying in the East and North if given winter protection.

◀ **'Nectarina'.** Origin: California. A medium-size, yellow-fleshed freestone with red blush and cavity. A low-chill variety with fruit ripening in mid-July. Good for the South and West.

◀ **'Southern Belle'.** Origin: California. A large, yellow-fleshed freestone with red blush. A low-chill variety that ripens in early August in California. Good for the South and West.

◀ **'Sunbonnet'.** Origin: California. A large, firm, yellow-fleshed clingstone with red blush. A moderate-chill variety (about 500 hours below 45°F), the fruit ripens in mid-July in California.

An assortment of pears, clockwise
from the upper left: 'Bartlett',
'Red Bartlett', 'El Dorado', 'Comice', 'Bosc',
'Anjou', 'Royal Vendee', 'Maxine', 'Forelle',
'Seckel', 'Orient', 'Old Home', 'Dr. De Portes'

A formal palmette espalier.
Pears on dwarfing rootstocks
are especially easy to train into
the formal espalier shapes.

PEARS

Pears, especially dwarf pears, are a fine
choice for the home gardener. The trees
are attractive even in winter, they require
little pruning after they begin to bear,
they begin to bear early, and the fruit
stores fairly well without any special requirements. The plants take well to formal or informal training, so space is no
problem.

Standard pears will spread 25 feet
across and grow as tall or taller. A dwarf
in natural shape needs a space about 15
feet square, but with the training methods
described in this book you can grow a
pear flat against a fence or wall, using
very little space.

A pear is as trainable as an apple, and a
trained tree can last 75 years. Plant pears
as espaliers, train them to 45-degree angles
for an informal hedge, or try them in tubs
as a single cordon or on a trellis.

Pears bear on long-lived spurs, much as
apples do. These spurs last a long time if
you're careful not to damage them when
picking the fruit. You don't need to thin
fruit, but if a very heavy crop sets, remove fruit that is damaged or very under-
sized. Thin a few weeks before harvest.

All pears need a pollinator. Use almost
any other pear. However, 'Bartlett' is a
poor pollinator for 'Seckel'.

The one real drawback with pears is
fireblight, but a home gardener can work
around it by choosing varieties wisely.
Fireblight is at its worst in spring, when
insects carry it from tree to tree. Resistant
plants are the best answer. Cut off any infected tissue well below the infection and
burn it. Other pests are codling moth,
mites, pear psylla, and pear slug. See pages
29–33.

Most fruits are best when picked ripe, or
nearly so. Pears are the exception. A tree-
ripened pear breaks down and turns soft
and brown at the core. Always harvest
pears when they have reached full size but
are still green and firm. Hold them in a
cool, dark place if you intend to eat them
within a few weeks. For longer storage,
refrigerate the harvested fruit, and remove
it from cold storage about a week before
you want to use it. Pears ripen faster if
they are held with other pears in a poorly
ventilated spot. For fast ripening, place
several in a plastic container.

Early Varieties

◀ **'Ayres'.** Origin: Tennessee. A cross of
'Garber' and 'Anjou', the fruit is golden
russet with a rose tint. The flesh is juicy,
sweet, and good for eating fresh or for canning. The tree is fireblight resistant. Good
for the South.

◀ **'Clapp's Favorite'.** This large yellow
fruit with red cheeks resembles 'Bartlett'.
The flesh is soft, sweet, and good both
for eating and canning. The tree is at-
tractively shaped and very productive but
highly susceptible to fireblight. Since it is
hardy, this variety is best in cold, late-
spring zones. Good for the North and
West. Widely available.

◀ **'Moonglow'.** Origin: Maryland. This
large, attractive fruit is soft and juicy with
a mild flavor. Use it for canning or eating
fresh. The tree is upright, vigorous, and
heavily spurred and begins bearing a good
crop when quite young. It resists fireblight, so it is good wherever the disease
is a severe problem. Good for all zones.
Widely available.

◀ **'Orient'.** Origin: California. This nearly
round fruit has firm flesh that makes it a

Left: 'Clapp's Favorite'

Center, top to bottom:
'Moonglow', 'Bartlett', 'Bosc'

Below: 'Max-red Bartlett'

good canner; however, the flavor is too mild for a good fresh pear. The tree produces moderate crops and resists fireblight. Good for the South.

◀ **'Starkrimson'.** An attractive red-skinned sport of 'Clapp's Favorite'. It does well in the West or North but is susceptible to fireblight. It has good quality fruit. Good for the West and North.

Midseason Varieties

◀ **'Bartlett'.** This familiar commercial pear is yellow, medium to large, and thin skinned. The flesh is very sweet and tender, fine for eating, and good for canning as well. The tree does not have especially good form and is subject to fireblight. It takes summer heat, provided there is adequate cold in winter. In cool climates it sets poorly without a pollinator (use any variety but 'Seckel'). Good for all zones. Widely available.

◀ **'Devoe'.** Origin: New York. The long fruit somewhat resembles 'Bosc', but the greenish-yellow to yellow color is closer to 'Bartlett'. Good for the South.

◀ **'Douglas'.** This variety is a hybrid of the sand pear. The smallish fruit is greenish-yellow with some acidity. The flesh is tender and contains grit cells. The tree is markedly resistant to fireblight, and although hardy, it does not need long winter chilling. Good for the South and West.

◀ **'Lincoln'.** Called by some "the most dependable pear for the Midwest," this variety bears large fruits abundantly. The tree is extremely hardy and blight resistant. Good for the North and South.

◀ **'Magness'.** Origin: Maryland. The medium-size oval fruit has a slightly russet color. The flesh is highly perfumed. The tree is vigorous and spreads widely for a pear. This variety needs a pollinator and produces small amounts of good quality fruit. It is highly resistant to fireblight. Good for the South and West.

◀ **'Maxine' ('Starking Delicious').** Origin: Ohio. This large and attractive fruit has firm, juicy, sweet white flesh. The tree is somewhat blight resistant. Good for the North and South.

◀ **'Max-Red Bartlett'.** Origin: Washington. A bud mutation of 'Bartlett', the fruit is cranberry red on the tree, changing to dull red when picked. The flesh is finer grained and sweeter than 'Bartlett'. The tree form resembles 'Bartlett', but shoots and leaves have the reddish tinge of the fruit. It is susceptible to blight and in cool climates needs a pollinator. Good for the West.

◀ **'Parker'.** Origin: Minnesota. This medium to large pear is yellow with a red blush. The flesh is white, juicy, and pleasantly sweet. The tree is upright, vigorous, and fairly hardy but susceptible to fireblight. Good for the North.

Late Midseason Varieties

◀ **'Anjou'.** A French pear originating in the mild area near the Loire, the fruit is large and green, with a stocky neck. The flesh has a mild flavor and is not especially juicy, but firm. It stores well and is good for eating fresh or for canning. The tree is upright and vigorous but susceptible to fireblight. It is not recommended for hot-summer areas. Good for the North and West. Widely available.

◀ **'Bosc'.** Origin: France. This long, narrow fruit has a heavy russet color. The flesh is firm, even crisp, with a heavy perfume that makes some people consider it among the very finest pears. It is good fresh or canned, and is fine for cooking. The tree is large and highly susceptible to fireblight. Good for the North and West. Widely available.

'Seckel'

Right: 'Chojuro'
Bottom: 'Ya Li'
Below: 'Comice'

◄ **'Duchess'.** Origin: France. This pear is greenish-yellow and very large. The flesh is fine textured and of good flavor. The tree is symmetrical and bears annually. Good for the North.

◄ **'Gorham'.** Origin: New York. Of excellent quality, this fruit strongly resembles 'Bartlett', but ripens later and can be stored longer. The tree is dense, upright, vigorous, and productive. Good for the North and South.

◄ **'Mericourt'.** Origin: Tennessee. This pear is green to yellow-green, sometimes blushed deep red and flecked with brown. The creamy white flesh is nearly grit free and is good fresh or for canning. A vigorous tree, it will withstand −23°F during full dormancy. It resists both fireblight and leaf spot. Good for the South.

◄ **'Patten'.** Origin: Louisiana. This large, juicy pear is particularly good fresh and fair for canning. Since the tree is especially hardy, it should be considered for the northern Mississippi Valley where 'Bartlett' and 'Anjou' fail. Good for the North.

◄ **'Seckel'.** This is a small, yellow-brown fruit that is not especially attractive but has the finest aroma and flavor of any home garden pear. Eat it fresh or use it whole for spiced preserves. The tree is highly productive and very fireblight resistant. It sets fruit best with a pollinator (any pear but 'Bartlett'). Good for all zones. Widely available.

Late Varieties

◄ **'Comice'.** Origin: France. This large, round fruit is green to yellow-green, with a tough skin. This sweet, aromatic, and juicy pear is the finest for eating but is not recommended for canning. The large vigorous tree is slow to bear and moderately susceptible to fireblight. It sets fruit better with a pollinator and should be grown only on dwarfing quince rootstock. This is the specialty of the Medford region in Oregon, but it does well in home gardens along the California coast. Good for the West.

◄ **'Dumont'.** Of European origin, this is a large pear with blushed yellow skin. The flesh is firm and juicy, with a sweet, rich flavor. It is one of the best winter pears. The tree tends to bear in alternate years, especially as it grows old. Good for the North.

◄ **'Kieffer'.** This sand pear hybrid has large yellow fruit that is often gritty and therefore poor for fresh use, but it keeps well in storage and is excellent for cooking and canning. The tree is especially recommended because of a high resistance to fireblight amounting to near immunity. It needs little winter chill but stands both cold and heat well, so its range is wide. Good for the East, North, South, and Midwest. Widely available.

ASIAN PEARS (APPLE PEARS)

This group of pears, native to Japan and China, was selected for size, shape, flavor, and lack of grittiness. The fruits are eaten firm like an apple, and they will keep in the refrigerator for 4 to 8 months without getting soft like a 'Bartlett' pear. The fruits have their own characteristic flavor, texture, and juiciness. All are fireblight resistant and need cross-pollination with any other pear that flowers at the same time.

They are best grown on Asian rootstocks, although they are very dwarf on French and quince rootstock. Fruit should be thinned to one fruit per spur and this is best done when fruits are ¾-inch in diameter, 6 to 7 weeks after bloom.

All Asian pears are good for the South and West and can be grown in the East. Good availability in West Coast areas.

Varieties

◄ **'Chojuro'.** A flat, russet-skinned variety with a strong flavor. It is very firm, stores a long time, and bears regularly every year.

◄ **'Kikusui'.** A flat, yellow pear with good texture and flavor. It is a very juicy, mild-flavored variety. Pick when the skin begins to turn yellow.

◄ **'Shenseiki'.** A flat yellow pear with good texture and flavor, this is the earliest maturing quality Asian pear and should be picked when the skin is yellow.

◄ **'Twentieth Century'.** A flat, green pear with fine flavor. It is the most popular Asian pear in California. It tends to bear in alternate years since it crops very heavily. Thin to one fruit per cluster.

◄ **'Ya Li'.** A pear-shaped fruit with fine texture and flavor. It blooms earlier than most other pears, so it must have an early pollinator like 'Tsu Li' or 'Seuri'. Thin for best size and annual bearing. An extremely low chilling requirement (300 hours below 45°F) means this apple pear will fruit in warm southern areas.

Below: Because the fruit stays on the tree long after the leaves have fallen, persimmons are often grown just for their oranmental value.

Right: 'Hachiya' persimmon
Bottom right: An oriental persimmon in late summer

PERSIMMONS

The persimmon belongs to the same family of plants as the ebony tree of southern Asia. The American persimmon, *Diospyros virginiana*, grows as a native from Connecticut to Kansas and southward, but it won't take the extreme cold of the northern plains or northern New England. It has small, edible fruits up to 2½ inches in diameter.

The large persimmon found in the market is the Oriental persimmon, *Diospyros kaki*, and its many varieties. It could be far more popular than it is if more gardeners realized the great value of both tree and fruit. The tree grows well in any well-drained soil and makes a fine medium-size shade tree with large leaves that turn a rich gold to orange-red in the fall. A heavy crop of orange fruit holds on until winter, decorating the bare branches. It can be grown in the southern states and on the West Coast.

Persimmon foliage is large and glossy, with leaves reaching 4 to 6 inches in length. The new spring leaves are bronze or reddish, and in fall they turn to shades of yellow, pink, and red. The 2-inch fruit hangs on into the first frosts and is orange with a red blush. Eat it when it softens, or use it as you would applesauce or bananas. If you want to store it, mash the soft pulp out of the skin for freezing, and discard the tough skin.

Use a persimmon tree as an attractive background plant in a shrub border, or in front of evergreens (where it shows off its leaves and fruit best). Since the persimmon grows slowly, it takes well to espalier training. Train it informally against a flat surface, or use a trellis to form a persimmon hedge. It will also grow well as a single lawn tree, but you'll have a problem in late fall when the soft fruit drops and squashes.

Persimmons are often allowed to grow naturally, forming globe-shape trees up to 25 or 30 feet high. They can be pruned back in spring to keep them smaller. Little pruning is necessary, however. Train the young tree to 3 widely spaced scaffolds and leave it alone thereafter, or control it by cutting each year to strong lateral branches, removing as much growth as necessary to maintain the size you want. In pruning an espalier, cut off enough of the previous year's growth to expose the most interesting lines of the plant.

Fruit is borne on new wood. On a naturally shaped tree it will set on the outer portion. Thinning fruit is unnecessary but will help keep lawn trees neater.

American persimmons are normally dioecious, meaning that some trees are male, producing pollen but no fruit, while others are female. You will need a female tree for fruit and a male close by for pollen. Plant both unless you have wild trees near your garden. An occasional improved variety has fruit crops without requiring a separate pollinator, but these are not yet generally available.

Oriental persimmons set fruit without pollination. The large fruits are 3 to 4 inches in diameter and are usually picked in October before the first frost. Store them in the refrigerator and eat only after they soften. Placing them in a bag with an apple will hasten the ripening process. 'Fuyu' is the one persimmon that is not astringent when firm. You do not need to soften it before eating.

Oriental persimmons stand winter temperatures to about 0°F, but they need only a short chill period (100–200 hours below 45°F) to fruit well in southern locations.

In the West, the persimmon has no serious pests. In the East, a flat-headed borer may attack the trunk, but it can be removed by hand.

Varieties

◄ **American Persimmons.** Good varieties include 'Early Golden', 'Garretson', 'Hicks', 'John Rick', and 'Juhl'.

◄ **Oriental Persimmons.** Good varieties include 'Chocolate', 'Fuyu', 'Tamopan', 'Tanenashi', and 'Hachiya'. 'Hachiya' is the popular large fruit sold commercially. 'Chocolate' has dark flesh around its seeds and is a type of persimmon rather than a variety.

Left: A standard Japanese plum. This type of plum grows best in the West.

Below: 'Stanley'

Japanese plums overgrow and overbear. Cut back the long whips as discussed under "Apricots," and thin fruit at thumbnail size, leaving 4 to 6 inches between remaining fruits.

Tree plums don't lend themselves to confinement, so use bush types if your space is limited. Use bush types as shrubby screens or try them in containers.

Brown rot is a major concern and requires summer spraying. See page 31. Bacterial leaf spot is a serious problem for most Japanese plums in the South and the East, but it is not a problem on other types of plums. Japanese types do best in the West; European types best in the East; bush types grow well in the South, Midwest, and North.

European Plums

These plums tend to be small, and most varieties are egg-shaped. The flesh is rather dry and very sweet. Prunes from these plums are the sweetest and easiest to dry. The plants are fairly hardy but some varieties do well where winter is mild. All varieties are self-pollinating, except for those noted.

Early Varieties

◀ **'Earliblue'.** This blue plum has tender, green-yellow flesh resembling 'Stanley', but softer. The tree is hardy but bears late. Production is moderate, but fine for the home garden. Use 'Mohawk' as a pollinator. It is best planted in the North, and ripens in mid- to late July in Michigan.

PLUMS

Of all the stone fruits, plums are the most varied. They range from hardy little cherry plums and sand cherries, to hybrids with the hardiness of natives, to sweet European plums and prunes, to sweet or tart Japanese plums. Plums bear for 10 to 15 years and sometimes more.

Standard plum trees take space. Expect your tree to fill an area 15 to 20 feet square. Bush and cherry plums can be used in smaller spaces and reach 6 feet or so, but may spread as wide or wider. A dwarfed European plum on Nanking cherry roots will reach 10 to 12 feet in height.

Most plums need a pollinator, although European kinds are generally partly or entirely self-fertile. Check the variety list for pollinators.

Plums bear on spurs along the older branches with the heaviest production on wood that is from 2 to 4 years old. 'Damson' and the sand cherry plums need little or no thinning. All the large-fruit Japanese types must be thinned 5 to 8 weeks after bloom. Thin fruit to 4 to 6 inches apart. European plums should have clusters thinned to 2 or 3 fruits per spur. The young trees should be pruned as discussed on page 41. Bush varieties need their oldest shoots trimmed off at ground level after about four years of bearing to encourage new growth.

Right: 'Italian Prune'
Center: 'Green Gage'
Bottom: 'Yellow Egg'

Below: Plum blossom

Midseason Varieties

◀ **'Damson'.** This old plum from Europe is derived from a different species than other European plums. The smallish blue fruits are best for jam, jelly, and preserves. Improved varieties include 'Blue Damson', 'French Damson', and 'Shropshire Damson'. The trees are small and self-pollinating, and fruit ripens at the end of August or in September. It is a late plum in the North. Good for all zones. Widely available.

◀ **'Green Gage' ('Reine Claude').** This is an old European variety. Greenish-yellow fruit has amber flesh and is good fresh, cooked, or preserved. Trees are medium size and self-pollinating. Fruit ripens in mid-July, later in the North. Good for all zones because it has a low chilling requirement and is cold hardy. Widely available.

◀ **'Stanley'.** Origin: New York. The most widely planted European plum in the East, Midwest, and South, this tree has large, dark blue fruit with firm, richly flavored yellow flesh. It bears heavily every year, is hardy into central Iowa, and is self-pollinating. Fruit ripens after mid-August, into September in northern regions. Good for the North. Widely available.

◀ **'Sugar'.** This very sweet, dark blue plum is fairly large and excellent for home drying and canning. The trees are self-pollinating and bear in alternate years, with light crops in off years. Fruit ripens after July 15. Good for all zones.

◀ **'Yellow Egg'.** This golden yellow plum has a thick skin and yellow flesh. The round-topped, vigorous tree is hardy and productive. In the West the tree is planted in Washington. It is self-pollinating, and the fruit ripens in late August. Good for the North and West.

Late Varieties

◀ **'Bluefre'.** This large blue freestone has yellow flesh. The trees are vigorous, self-pollinating, and bear young. Fruit hangs on well after ripening. It has some sensitivity to brown rot. Fruit ripens early in September. Good for the North.

◀ **'French Prune'.** The small fruit is red to purplish black and very sweet with a mild flavor. This is the main prune variety in California. The tree is large and long-lived, often surviving even after orchards have become housing developments. It is self-pollinating, and the fruit ripens in late August to September. Good for the South and West. Widely available in California.

◀ **'Italian Prune' ('Fellenberg').** This dark blue plum is very sweet and good for dessert, canning, or drying. It has been the major plum of the Washington-Oregon area. The fruit ripens in late August and September. Good for the South and West. Widely available.

◀ **'President'.** This large, dark blue fruit has amber flesh and ripens very late, after other plums. It lacks outstanding flavor, but use it for winter cooking or canning. Use another late European plum as a pollinator. Fruit ripens in Michigan at the end of September. Good for the North.

Japanese (Oriental) Plums

Japanese plums have relatively large, soft, and juicy fruit. The plants are the least hardy of the various kinds of plum, although selected varieties are grown in the milder northern regions. Taste one to test for ripeness before you harvest. Most varieties are self-pollinating, but all plums set fruit better with a pollinator.

Early Varieties

◀ **'Bruce'.** Origin: Texas. This large plum has red skin, red flesh, and good flavor. The fruit matures early. The tree bears young and heavily. Use 'Santa Rosa' as a pollinator. Fruit ripens in June. Good for the North and West.

◄ **'Early Golden'.** Origin: Canada. This medium-size, round plum is yellow and of fair quality. The stone is small and free. The tree is vigorous, outgrowing other varieties, but it has a tendency to bear in alternate years. Thin carefully. Pollinate with 'Shiro' or 'Burbank'. Fruit ripens in Michigan in mid-July. Good for the North.

◄ **'Santa Rosa'.** Origin: California. This popular large plum has deep crimson skin. Its flesh is purplish near the skin and yellow streaked with pink near the pit. It is good for dessert or canning. Use any early or midseason plum for improved pollination. Fruit ripens in California in mid-June, late in the North. Good for all zones. Widely available.

Early Midseason

◄ **'Abundance'.** Origin: California. This red-purple plum has tender yellow flesh that softens quickly. It is good for dessert or cooking. The tree tends to bear every other year. Use 'Methley' or 'Shiro' as a pollinator. Fruit ripens in Michigan in late July. Good for the North.

◄ **'Methley'.** This small to medium fruit is reddish purple with red flesh and excellent flavor. It ripens over a long period, needing several pickings. The tree is upright, with hardy flower buds. For better crops pollinate with 'Shiro' or 'Burbank'. Fruit ripens in Michigan in mid-July, earlier in the South. Good for the North. Widely available.

◄ **'Satsuma'.** Origin: California. This is a blood plum with red juice. The meaty fruit is small to medium, with a dull, dark red skin, red flesh, and a small pit. The flavor is mild and good. Use it for dessert or preserves. Use 'Santa Rosa' or 'Wickson' as a pollinator. Good for all zones. Widely available.

◄ **'Shiro'.** This medium to large plum is round and yellow and has a good flavor. Use it fresh or for cooking. The tree produces heavily. Use 'Early Golden', 'Methley', or 'Santa Rosa' as a pollinator. Fruit ripens in early July in California and the South, in late July in Michigan. Good for all zones. Widely available.

Midseason Varieties

◄ **'Burbank'.** Origin: California. This large red plum has amber flesh of excellent flavor. The trees are fairly small and somewhat drooping. Use the fruit for canning or dessert. Use 'Early Golden' or 'Santa Rosa' as a pollinator. Fruit ripens in early August in the Northwest and in mid-July in California. Good for all zones. Widely available.

◄ **'Duarte'.** The medium to large, dull red fruit has silvery markings. The flesh is deep red. Fruit keeps well and is tart when cooked. Use 'Santa Rosa' as a pollinator. Fruit ripens in late July. Good for all zones.

◄ **'Ozark Premier'.** Origin: Missouri. This extremely large red plum has yellow flesh. The trees are disease resistant, hardy, and productive. Fruit ripens early in August. Good for the North, Midwest, and South. Widely available.

Late Varieties

◄ **'Elephant Heart'.** Origin: California. This large, thick-skinned fruit is mottled purple and green. The flesh is blood red. Trees are strong and hardy. Fruit ripens over a long period. Use 'Santa Rosa' as a pollinator. Fruit ripens in late July or August. Good for all zones.

◄ **'Late Casselman' and 'Late Santa Rosa'.** These firm, late-ripening plums resemble regular 'Santa Rosa' in tree shape and appearance of fruit, but the fruit is sweeter and much firmer. They mature 6 weeks later than regular 'Santa Rosa'.

Hardy Plums

These plums were especially selected and bred for the coldest northern and Great Plains climates.

◄ **'Pipestone'.** Origin: Minnesota. This large red fruit has tough skin that is easy to peel. The flesh is yellow and of excellent quality but somewhat stringy. The tree is vigorous and hardy, performing reliably in cold regions. Use 'Toka' or 'Superior' as a pollinator.

◄ **'Superior'.** Origin: Minnesota. This large, conical red fruit with russet dots and heavy bloom has yellow, firm flesh excellent for eating fresh. The tree bears very young and prolifically. Use 'Toka' as a pollinator.

◄ **'Toka'.** Origin: Minnesota. This large, pointed fruit is medium red, and often described as apricot colored. The flesh is firm and yellow, with a rich, spicy flavor. The tree is a heavy producer, spreading, and medium size, but it may be short-lived. Use 'Superior' as a pollinator.

◄ **'Underwood'.** Origin: Minnesota. This very large, red, freestone plum has golden-yellow flesh that is somewhat stringy but of good dessert quality. Ripening extends over a long season from July. The tree is vigorous and among the most hardy. Use 'Superior' as a pollinator.

◄ **'Waneta'.** Origin: South Dakota. This is a large, reddish purple plum with yellow flesh. Use 'Superior' as a pollinator.

Left: Pomegranates are one of the most ornamental fruits, either in flower or in fruit. The red-orange flowers and fruit are in bright contrast to the glossy bright green leaves.

Below: The edible portion of the pomegranate is the tart, sweet seed sacs that will develop inside each of these young fruits.

prune as you like to shape the plants, but no pruning is necessary if you choose to let your tree grow naturally.

Blossoms form on the current year's growth, and as the fruit grows heavier, it pulls down the slender new branches, making a decorative weeping effect. Although the plant will stand drought, the fruit will split if a tree is allowed to dry out completely and is then watered. For a good crop, keep the moisture level even. Thinning is not necessary. An excess crop is very decorative if left on the tree. The trees are self-fertile, so even a single specimen will bear fruit.

Normally pomegranates have no pest or disease problems, but leaves can develop fungus diseases in humid climates.

POMEGRANATES

With its shiny leaves, fleshy orange flowers, and bright red fruit, the pomegranate is one of the most beautiful fruiting plants. The leaves have a reddish tint in spring and are bright yellow in fall, providing a background that makes the fruits especially attractive.

Pomegranates are often thought of as a tropical or desert fruit, but in fact they withstand winter temperatures down to about 10°F. While they do ripen their fruit best in very hot, arid climates, you can harvest edible fruit in cooler areas. They are ideal plants for the desert Southwest because they tolerate considerable drought. Rain or irrigation can cause pomegranates to split if either occurs close to harvest time.

Many people who admire the pomegranate as an ornamental shrub or tree seem to have trouble eating it. The edible portion is the juicy scarlet flesh around the abundant seeds. If you score the skin just down to these seeds in about six places, cutting from stem to flower, you can open the fruit and expose all the seeds at once. Pomegranates are good in fruit salads and make an excellent syrup when cooked with sugar and a little water. This syrup is sold commercially as grenadine.

You can grow pomegranates as fountain-shape shrubs or as single- or multiple-trunk trees. They reach about 10 to 12 feet tall under ideal conditions but often remain smaller. A shrub can spread from 6 to 8 feet across. Since they fruit on new growth, you can prune them back heavily without loss of flowers or fruit. You can

Varieties

◄ **'Wonderful'.** This is the most common fruiting pomegranate you're likely to see in the nursery.

Above: 'Wonderful'

BERRIES

Below: 'Earliblue' blueberries

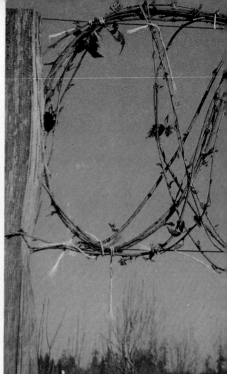

A little sunlight and a pot is all you need to grow luscious breakfast strawberries, and many of the other berries offer rich rewards for a small investment of time and space.

The small-fruited plants can return bumper crops with minimum effort and several of the shrubby or vining plants can also add beauty to the ornamental landscape.

In considering berries, you must work out the space and number of plants needed for a reasonable supply of fruit. If the plants are right for your climate and are given excellent care, the number of plants necessary to supply a family of five would be something like this:

Strawberries:	25 (20–30 quarts)
Raspberries:	24 (20–30 quarts)
Blackberries:	12 (9–15 quarts)
Blueberries:	6 (9–15 quarts)
Currants:	2 (6–12 quarts)
Gooseberries:	2 (6–12 quarts)

Strawberries are without question the easiest to work into any space you may have available. On a south-facing apartment terrace you can produce a crop in containers such as strawberry jars or moss-lined wire strawberry trees. An ideal plant for containers, where you can find it, is the European wild strawberry, or *fraises de bois*. This plant won't make runners. It grows in a clump, so a container planting stays compact.

The cane berries—blackberries and raspberries—take more space, although you can grow a few in large containers. If you train them carefully along a fence or trellis and keep them pruned, they won't take very much space; but they will produce heavy crops of fruit that you just can't buy, since the finest flavor disappears during transportation to your grocery store.

Blueberries, currants, and gooseberries make extremely ornamental shrubs, covered with bloom in spring, and with decorative fruit in later seasons. Blueberries require light, acid soil and constant moisture, so try them where you would grow azaleas. Currants and gooseberries are an interim host to a serious disease of five-needle pines, so in some areas you're not allowed to plant them. Where they are permitted, nothing takes less care, is more decorative, or gives a more useful crop.

Grapes, of course, are among the best landscaping plants, with lush foliage, fall color, and interesting vines. Use them on arbors, against walls, as fences, or as freestanding shrubs on a pole or trellis. Choose varieties recommended for your climate, since grapes are especially sensitive to heat.

BLACKBERRIES

Blackberries and raspberries are closely related and have similar growing requirements, but blackberries are larger and more vigorous, and some varieties are less hardy. Blackberries come in two fairly distinct forms—erect and trailing—and have a number of different names.

The ordinary blackberry is a stiff-caned, fairly hardy plant that can stand by itself if properly pruned. The trailing kind is generally called a *dewberry*, and it is tender and grown mainly in the South. In addition, trailing plants from the Pacific Coast are sold under their variety names; for example, 'Boysen' and 'Logan'. These varieties will freeze in the East and the North without winter protection.

Blackberries like a light, well-drained soil with high moisture-holding capacity. Do not plant them where tomatoes, potatoes, or eggplant have grown previously, since the site may be infected with verticillium wilt and the berries cannot grow there.

Plant in early spring a month before the last frost. Set plants 4 to 6 feet apart in rows 6 to 9 feet apart. Before planting, clip canes to 6-inch stubs and plant at the depth they grew in the nursery. As soon as new growth begins, cut any stubs that are left and burn them to protect plants from anthracnose, a fungal leaf spot disease that can infect bramble plants. It is a problem in moist, warm climates, especially in the South. Several inches of mulch will help keep soil moist, prevent weed growth, and help prevent suckers. Mulches such as fresh straw or sawdust will require added nitrogen. Use any high-nitrogen fertilizer

Left: Trailing blackberries need a trellis. See page 44.
Below: 'Ebony King' blackberries

at the rate of ½ to 1 pound per 100 square feet. Don't fertilize too heavily or you'll get lush plant growth at the expense of a fruit crop.

Blackberries fruit on twiggy side branches growing on canes of the previous season. The canes fruit only once and must be removed every year.

See pages 43–48 for pruning and training methods. The stiff-caned berries need no support, but can be confined between two wires to cut back on space. Dewberries should be cut to the ground after fruiting, and burned (if allowed in your area). The new growth of the last part of summer will fruit the following year, and burning the clippings reduces the chances of spreading disease.

If you disturb or cut roots of blackberries they will sucker badly. If you want more plants, chop off pieces of root beside the parent plants and set them in the new planting site like seed. If you don't want more plants, mulch the planting instead of cultivating for weed control. Blackberries can be more troublesome than any other cultivated plant and, if abandoned, can quickly grow out of control.

Blackberries are subject to enormous numbers of pests and diseases. Save yourself trouble by buying certified plants and keeping them away from any wild plants. A few varieties resist some diseases. Spray for blackberry mite, and don't worry too much about the rest.

Our list is divided into varieties for the South, the North, and the West, with the South and North separated into erect and trailing berries. Some varieties are recommended for more than one area of the country, in which case you will be referred to the main entry.

Varieties for the South
In much of the South, either dewberries or erect blackberries can be planted. In the warmest areas of Zone 3, dewberries are probably superior, although there are low-chill blackberries that do well. In Zones 5 and 6 choose only the erect blackberry, or be prepared to offer winter protection by burying canes under 2 inches of soil after the first frost, and then digging them out just as buds begin to swell.

Erect Blackberries
◀ **'Brainerd'.** Origin: Georgia. This large, high-quality fruit is excellent for processing. The plant is productive, vigorous, and hardy. Locally available.
◀ **'Brazos'.** This is a popular variety in Texas, Arkansas, and Louisiana. The large fruit matures early and bears over a long period. The plant is vigorous and somewhat resistant to disease. Locally available.
◀ **'Darrow'.** This hardy, heavy producer ripens in August from canes grown at the end of the previous summer. Fruit is large, glossy black, and of good quality. Widely available.
◀ **'Ebony King'.** Origin: Michigan. The large fruit is glossy black, sweet, and tangy. It ripens early and resists orange rust. Widely available.
◀ **'Eldorado'.** This very hardy and productive old variety resembles 'Ebony King' and is totally immune to orange rust.
◀ **'Flint'.** Origin: Georgia. This blackberry needs only moderate winter chill. The berries are fairly large in clusters of 8 to 15, and the plant is highly resistant to leaf spot and anthracnose. Locally available.
◀ **'Humble'.** This low-chill Texas variety has large, somewhat soft berries and comparatively few thorns. Locally available.

◀ **'Ranger'.** This large, firm berry is best when fully ripe. It is especially recommended for Virginia and similar climates.
◀ **'Smoothstem'.** Origin: Maryland. The berries ripen late and are rather soft. Production is quite heavy in large clusters. The plant is thornless and hardy from Maryland southward.
◀ **'Williams'.** Origin: North Carolina. The medium-size fruits ripen in late June and are very good fresh. The bush is semierect, vigorous, and thorny. It resists most cane and leaf diseases. Locally available.

Trailing Blackberries (Dewberries)
◀ **'Boysen' ('Nectar').** A Pacific Coast variety with large and aromatic fruit produced over a long season, this plant is vigorous and fairly thorny.
◀ **'Carolina'.** Origin: North Carolina. The plant is vigorous and productive, with very large berries. It resists leaf spot diseases. Locally available.
◀ **'Early June'.** Origin: Georgia. The large, round fruit is medium firm, has excellent flavor, and is acid enough for jam, jelly, and pies. Fruit ripens in early June. The plant is semithornless and somewhat resistant to anthracnose and leaf spot. Locally available.
◀ **'Flordagrand'.** Origin: Florida. The large fruit is very soft and tart, good for cooking and preserves. It ripens very early. Canes are evergreen. It must be planted with 'Oklawaha' for pollination. Locally available.
◀ **'Lavaca'.** This plant is a seedling of 'Boysen' that is hardier than the parent and more resistant to disease. The fruit is firmer and less acid. Locally available.

Right: 'Ollalie' blackberries

Below: 'Marion' blackberries

Below: 'Darrow' blackberries

◄ **'Lucretia'**. Origin: North Carolina. This hardy old favorite is vigorous and productive with very large, long, soft berries that ripen early. It needs winter protection in the North.

◄ **'Oklawaha'**. Origin: Florida. This plant resembles 'Flordagrand' and should be planted with it for pollination. Locally available.

◄ **'Young'**. Origin: Louisiana. This large, purplish-black fruit of excellent flavor is easy to pick. The plant produces few long canes. Anthracnose is a serious threat.

Varieties for the North
Trailing berries are too tender to grow in the North without special protection, but the fruit tends to be more flavorful than the erect kinds. Erect blackberries are not recommended for Zones 7 and 8 but will grow elsewhere, and may possibly succeed in the North if you bundle up the canes in straw and burlap for the winter.

Erect Blackberries
◄ **'Alfred'**. Origin: Michigan. This plant produces large, firm berries early. Locally available.

◄ **'Bailey'**. Origin: New York. The fruit is large, medium firm, and of good quality. The bush is reliably productive.

◄ **'Darrow'**. See "Varieties for the South."

◄ **'Ebony King'**. See "Varieties for the South."

◄ **'Eldorado'**. See "Varieties for the South."

◄ **'Hendrick'**. Origin: New York. The fruit is large, medium firm, and tart. The bush is reliably productive. Locally available.

◄ **'Raven'**. Origin: Maryland. This large berry is of high quality fresh or processed. The plant is erect, vigorous, and productive, but rather tender.

◄ **'Smoothstem'**. See "Varieties for the South."

◄ **'Thornfree'**. Origin: Maryland. The medium to large fruit is tart and good. The semiupright canes reach 8 feet, with up to 30 berries on each fruiting twig. The plant is rather tender. Widely available.

Trailing Blackberries
All of these berries are tender and need protection from cold.

◄ **'Lucretia'**. See "Varieties for the South."

◄ **'Thornless Boysen'**. This summer-bearing Pacific Coast berry is flavorful with a fine aroma, and grows on tender plants that must be trained. Bury the canes for the winter. Widely available.

◄ **'Thornless Logan'**. This Pacific Coast berry is acid, good for jam, pies, and for a syrup base for drinks. Bury the canes in winter.

Varieties for the West
The lists of varieties in this section have been divided into three groups: the Pacific Northwest, the interior, and California. There is a considerable amount of overlap among the three groups.

Berries for the Pacific Northwest like mild winters and cool summers. If you try them elsewhere, the climate should approach that mix, as it does along the northern California coast. California berries need a bit more summer heat and even milder winters. Berries for the interior are of a different kind, with stiff canes. They are hardier in cold winters.

These varieties are not limited to the region indicated, but they do well there. You will have more success with them if your climate resembles the one indicated.

Pacific Northwest
This heading refers to the mild coastal climates, not to the colder eastern areas.

◄ **'Aurora'**. This very early fruit is large, firm, and of excellent flavor. The canes are pliable and easy to train, and are most productive on the bottom 5 feet, so they do well planted closely and cut back heavily. Locally available.

◄ **'Boysen' ('Nectar')**. See "Varieties for the South."

◄ **'Cascade'**. The flavor of this berry is unsurpassed, fresh or preserved. The plant is vigorous and productive, but tender.

◄ **'Marion'**. The fruit of this midseason variety is medium to large, long, and of very good flavor. The plants send out a few vigorous canes that are up to 20 feet long and very thorny. This berry was the subject of an experiment by Arden Seets of Oregon proving that August-trained plants are more productive when planted only 2½ feet apart than with wider spacing.

◄ **'Thornless Evergreen'**. A top commercial berry in Oregon, this variety produces large, firm, sweet fruit. Plants are vigorous and produce heavily, but are very tender. Pinching canes at 24 inches encourages more canes and laterals, and may increase productivity. There is a thorny form.

◄ **'Thornless Logan'**. This large, reddish, tangy berry is best in pies or preserves. It is adapted to the Columbia River area of Washington, the Willamette Valley, and the central coast of California. There is a thorny form.

Right: 'Heritage' raspberries
Below: 'Ollalie' blackberries

The Interior

The following will grow in eastern Washington, and wherever cold is not too intense. They are stiff-caned, from rigid to somewhat trailing.

◄ **'Bailey'.** See "Varieties for the North."

◄ **'Darrow'.** Origin: New York. The berries are large and irregular, with firm flesh. They ripen over a very long season, sometimes into fall. The bush is hardy and reliable.

◄ **'Ebony King'.** See "Varieties for the South."

◄ **'Smoothstem'.** See "Varieties for the South."

California

The cool north coast suits the Northwest berries, too. Mountain and high desert gardeners may need berries from the Interior list.

◄ **'Boysen' ('Nectar').** See "Varieties for the South." In California, this variety provides an early crop from May 20 to June 20, depending on the area, and a second crop may extend the harvest through August. It is especially recommended for the Central Coast, the San Joaquin and Sacramento Valleys, and Southern California. Widely available.

◄ **'Himalaya'.** This late berry is grown in limited quantities in Northern California. Locally available.

◄ **'Ollalie'.** Origin: Oregon. This is the prime California variety, with large, firm high-quality berries that are shiny black, firm, and sweet. The canes are thorny and very productive. The plant has a low chilling requirement and resists verticillium wilt and mildew. It is especially good for Southern California.

◄ **'Young'.** See "Varieties for the South." This one does especially well in Southern California. It is similar to 'Boysen'.

RASPBERRIES

Raspberries are the hardiest of the cane berries, and perhaps the most worthwhile home garden crop for several reasons. First, prices for the market fruit are high, since care and labor are expensive. Then, too, market raspberries are subject to a long enough holding and handling period that fruit loses its finest flavor and may be bruised. Home garden fruit can be picked and eaten at its peak.

The thing that makes a raspberry a raspberry is the fact that it pulls free of its core when you pick it. Other bramble fruits take the core with them. The red raspberry is the most popular, but they come in a variety of colors and plant forms—red, yellow, purple, and black fruits, with the reds and yellows growing on trailing plants and the purple and black fruits growing on stiff plants. Do not try both red and black raspberries in the same garden. Reds sometimes carry a virus that they can tolerate but is fatal to blacks. Virus-free stock will spare you this trouble.

Unfortunately for southern gardeners, raspberries do poorly in much of the South. They need cold winters and a long, cool spring. Everbearing plants don't like any high heat. They can be grown by gardeners in Zones 5 and 6 and perhaps by a few high-ground gardeners in Zone 4 who are willing to risk some failures.

California and Arizona gardeners are similarly unfortunate. Raspberries do not like spring and summer heat, so only the red varieties will grow, and they are recommended only for coastal or mountain regions. The prime berry country in the coastal states is western Washington

around Puget Sound (Zone 8b), and the Willamette Valley of Oregon. Good berries grow through Zone 7, but may need winter protection in the coldest areas.

Wherever you grow them, cut the nursery plants to 6- to 12-inch stubs and plant them about 2 to 4 feet apart and 2 inches deeper than they grew in the nursery row. Your rows should be 6 or 7 feet apart.

On most raspberries, the fruit forms on side shoots along canes that grew the previous year. One group of red or yellow raspberries produces some fruit at the top of current-season canes in fall, then produces a second crop on the rest of the cane the following year.

New canes should be laid carefully along the rows until it's time to prune away old canes, then lifted and trained. Pull out suckers that sprout between the rows.

Raspberries are subject to all the same troubles as dewberries, but in the cold climates where raspberries will grow you'll have less trouble. Verticillium in the soil rules them out entirely, however. Black raspberries are susceptible to virus diseases and should be planted at least 700 feet from any reds.

If you want to enlarge a planting, it is important to know the difference between black and red raspberries. Blacks and purples arch their canes to the ground and root at the top to form new plants. If you want more plants, leave a few canes unpruned and in late summer pin the tip to the ground. Throw on a little soil if you like. Then dig and separate the new plant in spring.

Red raspberries send up root suckers. You can dig and replant them just before growth begins. Take a piece of root and cut back the top.

Left: 'Heritage' raspberries
Below: 'Latham' raspberries

Raspberries are extremely hardy, so no special protection is needed except in the coldest mountain and plains climates. Where winter temperatures stay extremely low for long periods, and winds add to the chill, you should protect your plants.

Lay canes of the current season along the row or trellis, pinning portions that arch upward. Be careful not to snap them. Where mice are not likely to be a problem, cover the canes with straw or sawdust to a depth of several inches, and then cover the mulch with poultry netting to hold it in place. If mouse damage is probable in winter, bury the canes in earth about 2 inches deep.

In spring, uncover the canes before they begin to leaf out, just as the buds swell. If the buds break while still covered, they will be extremely tender to even light frost.

Single-crop and everbearing raspberries may be red or yellow. They require trellising.

Single-crop Varieties

All fruit is borne on laterals that sprout from the year-old canes. There is one crop per season, either in late spring or early summer.

◀ **'Amber'**. This yellow berry is an excellent dessert fruit. Good for the North.

◀ **'Canby'**. These large, firm, midseason berries are good for freezing. The plants are semithornless and do best in light soils in the Northwest. Good for the West.

◀ **'Cuthbert'**. Once the leading commercial raspberry, and still unexcelled for dessert, canning, or freezing, this variety is low yielding and difficult to pick, but these are not problems in the home garden. Good for the West. Locally available.

◀ **'Fairview'**. These berries are large to fairly large and light red. The tall, branched canes are moderately hardy. It is especially suited to western Washington, and generally good for the West.

◀ **'Hilton'**. This berry is the largest of all reds, very attractive, and of excellent quality. Plants are vigorous, productive, and hardy. Good for the North.

◀ **'Latham'**. This early midseason variety is the standard eastern red raspberry. The berry is large, firm, and attractive, with a tart flavor. Plants are somewhat resistant to virus disease. Good for the South and West. Widely available.

◀ **'Newburgh'**. Often spelled 'Newberg', this productive midseason red variety yields large, firm berries. Good for the North and West.

◀ **'Pocahontas'**. This recent introduction has large, firm, medium-red berries with a tart flavor. The plant is winter hardy and productive. Good for the South.

◀ **'Puyallup'**. These late-ripening large berries are somewhat soft. The plant does best in light soils in the Northwest, and

Netting is often necessary to keep birds from eating berries.

'Blackhawk' black raspberries

is generally good for the West. Locally available.

◀ **'Summer'.** This medium to large berry is firm and sweet, with intense flavor. Some strains crumble badly. The plants do well in heavy soil and are recommended for western Washington or along the coast to Monterey, California. Locally available.

◀ **'Sunrise'.** This early variety offers firm, fine-textured fruit of good quality. The plant is hardy and very tolerant of anthracnose, leaf spot, and cane blight. Good for the South.

◀ **'Taylor'.** This variety offers mid- to late-season, attractive, firm red berries of excellent quality. Plants are vigorous and hardy. Good for the North. Locally available.

◀ **'Willamette'.** The berries ripen in midseason, are large, round, firm, and good for freezing or canning. This is a vigorous, widely planted commercial variety. Good for the West.

Everbearing (Summer and Fall Bearing) Varieties

Everbearing raspberries produce a crop in fall at the end of new canes, and then another in early summer of the following year. In California the second crop may not survive the heat. In the Northwest, these may produce a little fruit throughout the spring and summer.

◀ **'Cherokee'.** The berries are large and firm. The plant is winter hardy and productive. Good for the South, particularly the Piedmont area of Virginia.

◀ **'Durham'.** These berries have very good flavor. The plants are very hardy and productive, bearing a second crop early. Good for the North.

◀ **'Fallgold'.** Golden yellow berries distinguish this plant. Good for the North. Widely available.

◀ **'Fallred'.** The berries are of fair quality, but often crumbly. The plants are nearly thornless. The first crop appears in spring. Good for the South and North. Widely available.

◀ **'Heritage'.** Medium-size, firm fruit ripens in July and September. The plants are vigorous and stiff-caned and need little support. Mow all the canes in late winter to get a single August crop and to save pruning. Good for all zones. Widely available.

◀ **'Indian Summer'.** The fruit is large and of good quality. The first crop is light, the fall crop very late and abundant. Good for all zones.

◀ **'Ranere' ('St. Regis').** This variety has small, bright red berries of only fair quality, but provides a second crop, except in the warmest areas. Good for the West. Locally available.

◀ **'September'.** These medium to large berries are of good quality. The plant is vigorous and hardy—one of the best in the coldest regions. Good for the North and South.

◀ **'Southland'.** Recommended for farther south than any other, this berry was developed in North Carolina, but is not recommended for the coastal plain. It has large fruit of fair quality. Good for the South.

Purple Varieties

These raspberry plants are tall and stiff and bear a single crop.

◀ **'Amethyst'.** This early berry is a good-quality introduction from Iowa. Good for the North.

◀ **'Clyde'.** This early berry is large, firm, dark purple, and of excellent quality. The plant is vigorous. Good for the North.

◀ **'Sodus'.** This midseason berry is large, firm, and of good quality, but tart. Plants are productive. Good for the North.

Black (Blackcap) Varieties

Gardeners in the South and West should be aware that black raspberries are least able to tolerate mild climates. They need cold and do poorly in western Washington, although they are planted in the Willamette Valley and elsewhere in Oregon. They bear a single crop on 1-year-old canes.

◀ **'Allen'.** This variety provides large, attractive berries on a vigorous and productive plant. Good for the North.

◀ **'Black Hawk'.** This late variety bears large berries of good flavor and yield. Good for the North. Widely available.

◀ **'Bristol'.** These attractive, glossy black berries are large, firm, and of good quality. They must be fully ripe or you can't pick them. Good for the South and West. Widely available.

◀ **'Cumberland'.** This favored variety has large, firm berries of fine flavor. Plants are vigorous and productive. Good for the South and North. Widely available.

◀ **'Logan' ('New Logan').** This variety produces heavy crops of large, glossy, good-quality berries. Plants hold up in drought and tolerate mosaic and other raspberry diseases. Good for the South and North.

◀ **'Manteo'.** Origin: North Carolina. The fruit resembles 'Cumberland' but the plant survives farther south than any other. Good for the South. Locally available.

◀ **'Morrison'.** This variety bears later than others listed here. The berries are large, but the crop may be light. Good for the South and West.

◀ **'Munger'.** The medium-size fruit is of good quality. The plants are especially recommended for western Oregon. They are worth trying in western Washington, but may succumb to disease. Good for the West.

Right: Keep birds away from berries with plastic netting. See page 29.

Below: 'Earliblue' blueberries

BLUEBERRIES

Blueberries demand the right climate and planting soil, but take very little care if you provide suitable conditions. They are about as hardy as a peach, but need a fair amount of winter chill, and will not grow well in mild winter climates.

Blueberries belong to the heath family, and count azaleas, rhododendrons, mountain-laurel, and huckleberries among their cousins. If any of these grow naturally near your garden, or if you have prepared an artificial site that suits them, then blueberries will also do well.

Blueberries like soil rich in organic material such as peat—very acid, but extremely well drained. Such soils are usual in areas of high rainfall, which is lucky, since the berries need constant moisture, even though they cannot tolerate standing water.

There are major commercial plantings of blueberries in sandy soils in New Jersey, especially Burlington and Atlantic Counties; in Michigan, in certain areas of the Lower Peninsula; in Washington and Oregon; and to a certain extent in New York, Massachusetts, and Indiana.

Southern gardeners have a choice of two kinds of blueberries, depending on climate. The high-bush blueberry grows commercially in large plantings in southeastern and western North Carolina. A home gardener who hopes to succeed with the plant should live in, or north of, that area. If you know of native blueberries near your home, nursery plants should do well.

The rabbiteye blueberry, or southern high-bush blueberry, grows wild along streambeds in Georgia and northern Florida. With proper care it thrives where muscadine grapes succeed.

Soil must be extremely well drained and acid. Plant in raised areas if there is any chance of water standing around the roots for a day or more. For both drainage and acidification, add large amounts of peat moss or other organic material to the planting soil, up to three-quarters peat moss by volume for soils that tend to be heavy. Never add manure; it is alkaline. Dig a planting hole somewhat broader and deeper than the roots of the young plant. Never cramp the roots into a small hole. Spread the roots in the hole and press soil firmly over them.

Set high-bush blueberry plants about 4 feet apart. Choose two varieties for cross-pollination. Since the rabbiteye plants grow much larger, you can set them up to 8 feet apart, although they can also be set closer and blended into each other.

Do not feed plants the first year. In succeeding years, use cottonseed meal, ammonium sulfate, or any product suitable for camellias, azaleas, or rhododendrons.

Blueberries require constant light moisture in the soil, and cultivating damages their shallow roots. For both these reasons, you should mulch the plants heavily. Use any organic material such as straw, leaves, peat moss, or a combination, and renew it regularly to keep it about 6 inches deep. Some materials will use nitrogen as they decay, so you will have to compensate with extra feeding.

Pruning is similar for both kinds of blueberries. Leave them alone for two or three seasons, trimming only tangles or broken twigs. Then, to cut back a little on the extremely heavy crops of small berries, remove some of the oldest canes or stems, and clip out a few of the fruiting twigs. If you do nothing, you will still get fruit, but it will be small and eventually will decline in quantity.

Blueberries suffer from very few difficulties, but birds will take them all unless you net the plants. Nurseries carry suitable netting.

Always taste blueberries before picking. Some look fully mature when still quite acid.

Approximately the same varieties are used throughout the country, since the conditions for growing them are so similar.

Early Varieties

◀ **'Earliblue'.** Origin: New Jersey. One of the best for all areas, this berry is large, light blue, and firm. The picking scar is small, so fruit keeps well and resists cracking. Plants are upright and comparatively hardy. Good for all zones. Widely available.

Below: 'Northland' blueberries

Blueberry varieties. From top to bottom: 'Dixi', 'Blueray', 'Coville', 'Bluecrop'

◀ **'Ivanhoe'.** One of the best berries, this is large, light blue, and firm. The plant is very tender. Good for the South.

◀ **'Northland'.** Origin: Michigan. This is a hardy variety. The fruit is medium size, round, moderately firm, and medium blue. The flavor is good. The plant is spreading but reaches only 4 feet at maturity. Good for the North and West. Locally available.

◀ **'Weymouth'.** Origin: New Jersey. The large, round berry has a dark blue skin and little aroma. It ripens very early and is best for cooking. The bush is erect, spreading, and very productive, but not vigorous. Good for the North and West.

Midseason Varieties

◀ **'Berkeley'.** Origin: New Jersey. This large, firm berry is pale blue and resists cracking. The bush is fairly upright and moderately hardy. Good for all zones and especially for the West Coast into Northern California. Widely available.

◀ **'Bluecrop'.** Origin: New Jersey. The fruit is large, light blue, and rather tart, but stores well and is good for cooking. The berries stand cold well, which makes the plant good for the shortest Michigan growing seasons. The plant is upright and medium hardy. Widely available.

◀ **'Blueray'.** Origin: New Jersey. The fruit is very large, firm, and sweet. The plant is upright and spreading. Good for all zones, and especially recommended for Washington. Widely available.

◀ **'Collins'.** Origin: New Jersey. The fruit is large, light blue, firm, and sweet. It resists cracking. The plants are erect, well shaped, and fairly hardy, but not consistent in production. Good for the North and West.

◀ **'Croatan'.** The fruit is medium size and quick to ripen in warm weather. The plant is canker resistant. Good for the South, and especially recommended for North Carolina. Locally available.

◀ **'Stanley'.** Origin: New Jersey. This is a widely recommended variety. The fruit is medium size and firm, with good color and flavor. The bush is hardy, vigorous, and upright. Since there are few main branches, pruning is easy. Good for the North and West.

Late Varieties

◀ **'Coville'.** Origin: New Jersey. This is an inconsistent variety with large, light blue fruit that remains tart until near harvest. The plant is medium hardy. Good for all zones. Widely available.

◀ **'Delite'.** This is the only variety that develops some sugar early, so picking is easier. The berries are medium large and may be reddish under the bloom.

◀ **'Dixi'.** Origin: New Jersey. The name is not an affectionate term for the South, but Latin for "I have spoken" or, loosely, "That's my last word." It was given by the developer, F.V. Coville, on his retirement. The fruit is large, aromatic, flavorful, and good fresh.

◀ **'Menditoo'.** This dark blue berry is large, medium firm, and sweet. The bush is fairly vigorous and spreading. Fruit ripens over a long period, which is convenient for home gardeners. Locally available.

◀ **'Southland'.** The firm, light blue berries have a waxy bloom, and may have tough skin late in the season. This is a good Gulf Coast plant. Locally available.

Right: 'Red Lake' currants

Below: Red currants make delicious and beautiful jelly.

Below: 'White Imperial' currants
Bottom: Gooseberries

CURRANTS AND GOOSEBERRIES

Currants and gooseberries are among the most beautiful of the small fruits, but they are good home garden shrubs for other reasons as well.

You won't often see fresh fruit in the market, since crops from the limited commercial plantings go to processors for commercial jellies and canned fruits. But since the plants are ornamental, easy to care for, and productive, northern gardeners can tuck a few among other shrubs for the bloom, fruit, and fall color. The crop can be used for jelly, pie, or just fresh eating for those who like a tart fruit.

We discuss only the red and white currants of the species *Ribes sativum* and the gooseberries *Ribes uva-crispa* and *R. hirtellum*. The black currant, *Ribes nigrum*, so aromatic and rich in vitamin C, is unfortunately banned almost everywhere, since it is part of a disease cycle of five-needle pines. Spores of white pine blister rust from miles away spend part of their lives on the currants, then transfer to pines growing within about 300 feet. Currants are banned where there is a white pine timber crop, but check your own and neighbors' gardens for pines with bundles of five needles, and if you find them, don't plant currants. The other *Ribes* species also take part in transferring this disease, and they, too, are banned in some areas. Ask at your nursery, and do not transport or plant any currant or gooseberry from outside your region without checking with your Cooperative Extension Office.

Fall or winter planting is a good idea, since the plants leaf out early. In cold climates, plant right after the leaves drop and the roots will be established before winter. Space the plants about 4 feet apart, or set them closer if more convenient, but expect them to grow less vigorously. If summer is hot, plant them against a north wall. In most areas, plant in the open, but be sure soil moisture is constant. Set the plants a little deeper than they grew in the nursery.

Currant Varieties

◀ **'Perfection'.** This old variety has medium-size red fruit in loose clusters. The plant has good foliage and is upright, vigorous, and productive. Good for Washington and Oregon. Widely available.

◀ **'Red Lake'.** Origin: Minnesota. Recommended everywhere that currants will grow, this variety yields medium to large, light red berries in long, easy-to-pick clusters. The plants are slightly spreading. They produced the highest yield in Canadian trials and also produce well in California. Widely available.

◀ **'Stephens No. 9'.** Origin: Ontario, Canada. This is a good Great Lakes variety, with fairly large, medium red berries in medium clusters. Plants are spreading and productive. Locally available.

◀ **'White Grape'.** This is a white variety that is widely sold, but perhaps surpassed in quality by 'White Imperial', if you can find it.

◀ **'Wilder'.** This very old variety from Indiana yields dark red berries that are firm but tender and very tart. Plants are large, hardy, and long-lived.

Gooseberry Varieties

◀ **'Clark'.** Origin: Ontario, Canada. This fruit is large and red when ripe. Plants are usually free of mildew. This is a good Canadian variety.

◀ **'Fredonia'.** Origin: New York. The large fruit is dark red when ripe. Plants are productive and vigorous, with open growth.

◀ **'Oregon Champion'.** Origin: Oregon. The medium-size fruit is green. This is a good variety for all West Coast growing areas.

◀ **'Pixwell'.** Origin: North Dakota. This is a very hardy variety for central and plains states. The berries hang away from the plant, making them easy to pick, and the canes have few thorns. Widely available.

◀ **'Poorman'.** An American variety with red fruit, the plants are spiny and spreading. Good for the Pacific northwest and the central states.

Wine grapes can be trained to a head; that is, they are freestanding, with the fruiting stubs, or spurs, selected early — about 4 per vine — then clipped back each year until the leafless plant looks like a caveman's club.

Muscadines and bunch grapes differ in pruning once established. See page 49.

Feeding
Grapes need be fed only nitrogen, and may not always need that. If the leaves yellow and there is little growth in the early part of the season, they definitely need feeding. If you're not sure, try a feeding to see the result. Late feeding during the ripening period can force excessive growth and spoil the fruit.

Harvest
Harvest grapes by taste and appearance. When you think the bunch looks ripe, taste a grape near the tip. If it's good, cut the bunch.

Sometimes grapes never taste sweet, no matter how long you wait. That simply means that you have planted the wrong variety for your area. Either switch to another variety, or replant the one you stubbornly insist on in a hot spot against a south wall or in a westward-facing corner.

If vines are overcropped with too many bunches of grapes the grapes will never get sweet. This can be remedied in future years by more extreme pruning in the dormant season or by thinning the grape bunches to balance the leaf area with the grape berry load.

GRAPES
In the earliest periods of human history, four foods were recognizably important. In the North there were apples and honey. In the South there were olives and grapes.

The American grape entered our history more recently than the vine of Europe, but it has already played an important role, since its roots saved the European grape from extinction during the Phylloxera plague of the last century. This plague threatened to destroy the European grapes and the only remedy was grafting these grapes to American rootstocks. More recently, American grapes have entered into sturdy hybrids that carry European wine grapes far north of their original climate area.

Soil and Site
Grapes send their roots deep where they can, and they prefer a soil that is rich in organic material. You can encourage growth by adding an organic supplement at planting time and mulching the roots afterward. The site should have good air circulation, because grapes are subject to disease where air is stagnant.

Planting, Training, and Pruning
The two kinds of grapes are pruned differently. (See pages 45 and 49 for sketches and instructions.) In general, American grapes need cane pruning (long canes with 10 or more buds on each remain to fruit after pruning); and European grapes need spur pruning (permanent arms or branches are developed, and canes along these are cut back to two buds). There are some exceptions, however.

Paper bags over ripening grapes keep the birds from harvesting them early.

Center, from top to bottom: 'Freedonia', 'New York Muscat', 'Seneca'

Below: 'Van Buren'

Pests and Diseases

Grapes mildew badly, and need good air circulation and treatment with a fungicide. The classic remedy is copper sulfate. A number of pests attack grapes, especially certain beetles. Birds love grapes, but you can save the fruit by placing whole bunches in paper bags.

Varieties for the Northeast and Midwest

The American grapes are listed first, with a note when they are choice juice or wine grapes. French hybrids are listed second.

American Varieties

◄ **'Catawba'.** Origin: North Carolina. Good for wine or juice, this red grape is a popular commerical variety. It requires a long season to ripen and will do well in southerly areas with the longest growing seasons. Thinning will hasten development. Widely available.

◄ **'Cayuga White'.** This variety bears white grapes in tight clusters. They are of good dessert quality.

◄ **'Concord'.** Origin: Massachusetts. This late grape is so well known and widely planted that it hardly needs description. Often the standard of quality in judging American grapes, the dark blue slipskin berries are rich in the characteristic "foxy" flavor, which is retained after processing. Widely available.

◄ **'Delaware'.** Origin: New Jersey. The clusters and berries of this major wine grape are small, good for wine and juice, and excellent for dessert eating. Vines mildew.

◄ **'Fredonia'.** Origin: New York. This variety should be allowed to set heavily, as it sometimes has difficulty with pollination. This is the top black grape in its season. Vines are hardy. Widely available.

◄ **'Himrod'.** Origin: New York. This is the top white seedless grape throughout the northern states. 'Thompson' types replace it where weather is warmer. The vines are brittle and only moderately hardy. Widely available.

◄ **'New York Muscat'.** Origin: New York. Good for wine and juice, the reddish-black berries in medium clusters have muscat aroma, which is rich and fruity, not "foxy". Temperatures below −15°F can cause winter injury.

◄ **'Niagara'.** Good for wine and juice, this is the most widely planted white grape and more productive than 'Concord.' It is vigorous and moderately hardy. Widely available.

◄ **'Schuyler'.** Origin: New York. This grape resembles European grapes in flavor. It is soft and juicy with a tough skin. The vines are fairly hardy and disease resistant.

◄ **'Seneca'.** Origin: New York. The small to medium berries resemble European grapes, with tender golden skin and sweet, aromatic flavor. The vine is hardy and takes cane pruning, although one parent is a European type.

◄ **'Sheridan'.** Origin: New York. Large berries have tough black skin and require a long season to mature. It is a 'Concord' type and its vines are hardy.

◄ **'Steuben'.** Origin: New York. Blue-black berries grow in medium clusters and are sweet and aromatic. The vine is productive, hardy, and disease resistant. Widely available.

◄ **'Van Buren'.** Origin: New York. The small to medium clusters of jet-black grapes have a sweet, "foxy" flavor and are very good fresh. This is the earliest of the 'Concord' types: However, the juice holds its flavor less well than 'Concord'. The vine is hardy but subject to mildew.

Below: 'Concord'

Center, from top to
bottom: 'De Chaunac',
'Chancellor', 'Foch'

Below: 'Baco #1'

◀ **'Veesport'.** Origin: Ontario, Canada. Borne in medium clusters, these black grapes are good for wine and juice, and acceptable for fresh eating. The vine is vigorous.

French Hybrids

Spur prune these. They are hybrids of European and less well-known American grapes (not the 'Concord' type). All are for wine or juice, but are also good fresh.

◀ **'Aurora' (Seibel 5279).** This very early white grape is soft, with a pleasant flavor. It is a dependable producer and a vigorous grower, although it is better in sandy than in heavy soils. Choose it if early ripening is needed. Widely available.

◀ **'Baco #1' ('Baco Noir').** This mid-season variety produces small clusters of small black grapes. It is extremely vigorous and productive, but it tends to bud out early and is subject to frost injury. This is not a cold hardy variety. Widely available.

◀ **'Chancellor' (Seibel 7053).** This vigorous blue-fruited grape ripens in late midseason. The vines produce well even though they are subject to downy mildew.

◀ **'De Chaunac' (Seibel 9549).** This mid-season variety is one of the best blues, especially for home winemaking. The clusters are medium to large on vigorous and productive vines.

◀ **'Foch'.** This very early black variety bears medium-size clusters, with medium berries. Birds love them, so protect the crop. Vines are very vigorous.

◀ **'Seyve Villard 12– 375'.** Good for table use as well as for wine, these white berries ripen in late midseason. The vines are vigorous, productive, and hardy.

◀ **'Seyve Villard 5276'.** A midseason variety with large, compact bunches of white berries. It is moderately vigorous and subject to downy mildew and black rot.

Varieties for the West

The West is grape country, wherever you go, and yet many gardeners are disappointed in the fruit they harvest from their vines. The problem is usually a poor choice of varieties. Grapes, more than any other fruit, require the right climate and amount of heat to produce well. Too many gardeners buy vines because they like the fruit in the market, or because they know a famous name.

In general, western grape climates are divided into three groups. The first includes all of the West except California and the southwestern desert. Gardeners in these cool regions should choose an American grape of the "foxy"-flavored species, *Vitis labrusca.* The most highly recommended appear below under the heading, "Northwest". 'Concord', often sold by nurseries in the cool regions, is not successful in western Washington and Oregon. It requires more heat.

In California, the cooler coastal areas and coastal valleys are suited to American grapes and selected European varieties with a low-heat requirement. 'Concord' does well, but the popular 'Thompson Seedless' will almost always disappoint the home gardener. 'Perlette' is similar, but it was developed for the low heat of this climate. The inland Northwest and parts of Utah, Colorado, and Idaho can also use 'Concord' and 'Niagara' from the coastal California list.

In the hot inner valleys of the California coast range, there are major commercial vineyards growing all the renowned European wine grapes. The Napa-Sonoma wine region is well known, but there are also many wine grapes grown in newer plantings in southern Santa Clara County, San Benito County near Salinas, and north of Santa Barbara.

Below: 'Interlaken' Seedless

Left: 'Buffalo'
Bottom: 'Schuyler'
Below: 'Perlette'

The hot Central Valley climate is perfect for the European table grapes that you see on your grocer's counters. 'Thompson', 'Ribier', and 'Emperor' all do well.

The low and high deserts are not good grape country. The earliest maturing European varieties stand the best chance of producing a worthwhile crop.

Northwest Varieties

◄ **'Buffalo'.** Origin: New York. This grape ripens in midseason. It is a good grape for wine or juice with fairly large clusters of reddish-black berries. Cane prune this vigorous vine.

◄ **'Interlaken Seedless'.** Origin: New York. This grape ripens early, with medium clusters of small, seedless berries with greenish-white skin that adheres. The flesh is crisp and sweet. The grape resembles 'Thompson Seedless', but has more interesting flavor overtones. The vine is fairly hardy and best with cane pruning. Widely available.

◄ **'Ontario'.** Origin: Canada. These white berries form in fairly loose clusters. The vines are vigorous, productive, fairly hardy, and prefer quite heavy soils. Cane pruning is best.

◄ **'Schuyler'.** See "Varieties for the North."

◄ **'Seneca'.** See "Varieties for the North."

◄ **'Van Buren'.** See "Varieties for the North."

California Coast Varieties

◄ **'Cardinal'.** Origin: California. These large, dark red berries ripen early and have firm, greenish flesh. The medium-size clusters are extremely abundant, up to three per cane, but irregular. The vine can run rampant. Use it to cover an arbor or summerhouse. Spur pruning is best.

◄ **'Concord'.** This grape, described earlier,

does not like high California heat or the coolest Northwest summers, but does well anywhere in between. Cane pruning is best. Widely available.

◄ **'Delight'.** Origin: California. This grape ripens early, yielding well-filled clusters of large, greenish-yellow berries with firm flesh and a distinct muscatlike flavor. Spur pruning is best. Locally available.

◄ **'Niabell'.** Origin: California. This midseason variety produces well-filled clusters of large, black berries good fresh or as juice. Vines are vigorous, resist powdery mildew, and can be pruned to long canes. Cane pruning is best.

◄ **'Niagara'.** This variety, described earlier, ripens midseason to late midseason. Cane pruning is best. Widely available.

◄ **'Perlette'.** Origin: California. This early variety bears large, compact clusters of small white to yellowish grapes that are very tender and juicy with mild flavor. They tolerate high temperatures, but also ripen where overall heat is fairly low. The vine is vigorous, very fruitful, and needs thinning. Spur pruning is best. Locally available.

◄ **'Pierce'.** If 'Van Buren' is the cool-summer 'Concord', then this is the hot-summer 'Concord'. Grow it in the warmer regions of central California where you want a "foxy" black slipskin. The vine is very vigorous. Cane pruning is best. Locally available.

California Valley (hot climate) Varieties

◄ **'Cardinal'.** See "California Coast Varieties."

◄ **'Emperor'.** This late-ripening, large, red grape has flesh so firm it seems to crunch. It is adapted to the hottest part of the San Joaquin Valley. The berries are firm and will store longer than other varieties. It is unsatisfactory in cooler areas. Spur pruning is best.

◄ **'Muscat of Alexandria'.** These late midseason, large, green berries splotched with amber and in loose clusters are not pretty, but have an unparalleled musky, rich flavor. They lose flavor if held too long, so are best eaten fresh from the home garden. These grapes can also be dried as seeded raisins. Spur pruning is best. This variety requires the moderately high heat of the San Joaquin Valley or other inland valleys but not the desert. Muscats are often used to make sweet dessert wine. Unfortified muscat wine is a treat with desserts or fruit.

◄ **'Niabell'.** See "California Coast Varieties." This is a juice grape. Locally available.

◄ **'Perlette'.** See "California Coast Varieties." Try this one in low desert regions as it resists sunscald. Locally available.

◄ **'Red Malaga'.** This early midseason variety bears large clusters filled with large pink to reddish-purple berries that have little flavor. Use long spur pruning or cane pruning with thinning of flowers. Locally available.

Left, from top to bottom:
'Zinfandel', 'Chardonay', 'Pinot Noir'

Below: 'Tokay'

◄ **'Ribier'.** This is a beautiful, early mid-season dessert grape with large, jet-black berries. They tend to soften quickly in storage and lose their mild flavor. The vines are overproductive. Use short spur pruning and thin the flowers.

◄ **'Scarlet'.** The compact clusters of this midseason variety hold jet-black berries with abundant, bright red juice that is sweet and richly flavored. It also has a mild "foxy" flavor. Vines are vigorous and ideal for arbors. Leaves turn dark red in fall. Cane pruning is best. Locally available.

◄ **'Thompson Seedless'.** This is the top commercial seedless green grape, ripening in early midseason. Clusters are well filled with rather long fruit; and the flavor is mild. These are excellent as a fresh grape if clusters are thinned. They are also used for raisins. Grow only in hot climates. (Try 'Perlette' or 'Delight' if in doubt.) Cane pruning is required.

◄ **'Tokay'.** This late midseason variety bears large clusters of large, very firm red grapes that are attractive but have little flavor. It does well in the Lodi area, and the cooler valley climates. Use 'Emperor' in hotter climates. Spur pruning is best. Locally available.

Wine Grapes for the West
The list includes three each of the best-known red and white grapes. They change character over short distances, so unless you know that professionals grow them near you, don't count on getting the best quality.

◄ **'Cabernet Sauvignon'.** This is the great European black grape used to make the red Bordeaux wines of France. Cane prune for best results.

◄ **'Chardonnay'.** This popular white grape is used to make the famous French White Burgundy. It is a vigorous grower and moderate producer. The clusters of berries are small. It is best in cool coastal areas and should be cane pruned.

◄ **'Chenin Blanc'.** The vines on this white grape variety are vigorous and productive. It yields medium-size berries and clusters. The coastal valleys and the San Joaquin Valley have the best climates for this variety. It should be cane pruned.

◄ **'French Colombard'.** This productive white variety yields a grape that is high in acid. It is adapted to coastal valleys and the Central Valley of California and bears medium-size berries and clusters. It can be cane or spur pruned.

◄ **'Pinot Noir'.** This small black grape is used to make the French Burgundy wines. It is difficult to handle but should be cane pruned.

◄ **'Zinfandel'.** This is a California specialty for red wine. You can probably grow this better than any other, as it seems to make drinkable wine in a variety of climates.

Right, from top to bottom:
Muscadine grapes, 'Golden Muscat', 'Stuben'

Below: 'Catawba'

Varieties for the South

In the South, two quite different types of grapes are widely grown, both of American origin. The bunch grape is typified by 'Concord'. Although this type prefers a cool climate, varieties are available for most regions. The real southern grape is, of course, the muscadine, with its smaller clusters of berries and liking for Cotton Belt weather.

Bunch Grape Varieties

◀ **'Catawba'.** This grape, described in "Varieties for the Northeast and Midwest," is especially recommended for South Carolina. Widely available.

◀ **'Champanel'.** These large, purple grapes grow in clusters that are long but sometimes poorly filled. The berries are large and juicy, with high acidity. Vines are vigorous and productive. Locally available.

◀ **'Concord'.** See "Varieties for the Northeast and Midwest."

◀ **'Delaware'.** See "Varieties for the Northeast and Midwest."

◀ **'Fredonia'.** See "Varieties for the Northeast and Midwest."

◀ **'Golden Muscat'.** Origin: New York. These large, golden berries are borne in large clusters. The flavor is typically rich and fruity—muscat, not "foxy". Vines are tender. It is especially recommended for Georgia.

◀ **'Himrod'.** See "Varieties for the Northeast and Midwest."

◀ **'Niagara'.** See "Varieties for the Northeast and Midwest."

◀ **'Seneca'.** See "Varieties for the Northeast and Midwest."

◀ **'Steuben'.** See "Varieties for the Northeast and Midwest."

Muscadine Varieties

Many muscadines are sterile and need a pollinator. The varieties described below as "perfect" will pollinate themselves and any other variety.

◀ **'Hunt'.** Origin: Georgia. This dull black fruit is unusual in ripening evenly. The quality is excellent, high in sugar, and very good for wine and juice. The vine is vigorous and productive. This variety is unanimously recommended for home and commercial planting by the Muscadine Grape Committee.

◀ **'Jumbo'.** This is a very large black muscadine of good quality. It ripens over several weeks, so is excellent for fresh home use. Vines are disease resistant.

◀ **'Magoon'.** Origin: Mississippi. Perfect. Reddish purple berries are medium size and have a sprightly, aromatic flavor. The vine is productive and vigorous.

◀ **'Scuppernong'.** Origin: North Carolina. This is the true muscadine, the earliest variety named. Most people call any similar grape a scuppernong, but this is the real variety. Fruit color varies from greenish to reddish bronze, depending on sun. It is late-ripening, sweet, and juicy, with aromatic flavor. Good for eating fresh or for wine or juice.

◀ **'Southland'.** Origin: Mississippi. Perfect. This very large grape is purple and dull skinned, with good flavor and high sugar content. The vine is moderately vigorous and productive. Good for the central and southern portions of the Gulf Coast states.

◀ **'Thomas'.** This standard grape has reddish black, small to medium berries that are very sweet and excellent for fresh juice. Locally available.

◀ **'Topsail'.** Origin: North Carolina. Clusters of three to five berries have green fruit splotched with bronze. This is the sweetest of all muscadines and very good for fresh use. It is often a poor producer because it does not form perfect flowers. Vines are not very hardy, but are disease resistant.

◀ **'Yuga'.** Origin: Georgia. These reddish bronze berries are sweet and of excellent quality, but ripen late and irregularly. They are fine for home gardens.

Left: Botanically, the strawberry "fruit" is a receptacle; the true fruits are the "seeds" imbedded in its surface.

Below: This strawberry "tree" is made by planting in pockets in a wire column. The column is lined with moss and filled with soil.

STRAWBERRIES

If you have grown strawberries for any length of time, you know that flavor and yield are not predictable, but vary from year to year depending on spring growing conditions. Also, if you have gardened in several locations you know that the best variety in one place may be only fair in another. A good nursery can be a big help, since the staff keeps abreast of developments in plant breeding and offers plants that should succeed. Your county agricultural agent can help, too, especially if you've had trouble in other seasons.

Planting and Care

Strawberries can be grown in either the matted row or the hill system. There are two types of matted rows. In one, all runners are allowed to grow; in the other, only the earliest to form remain, spaced about 8 inches apart. The latter, spaced-runner system gives you larger berries, easier picking, and larger total yield. See the illustration for the arrangement of runners. The rows should be spaced 3 to 4 feet apart when you plant.

In the hill system, plants are 12 inches apart in the row and all runners are picked off. The rows are spaced about 12 to 15 inches apart in groups of three. Each group is separated by an aisle 24 to 30 inches wide so that you can walk among the plants to pick or care for them. The system lends itself to everbearing strawberries, or single-crop kinds that don't send out many runners.

To encourage vigorous growth of regular varieties, remove blossoms that appear the year the plants are set out. The year that everbearing kinds are planted, remove all blossoms until the middle of July. The later blossoms will produce a late-summer and fall crop.

Plant strawberries in soil with good drainage, and mound the planting site if you're not sure about the drainage. See the illustration for proper planting depth. The new leaf bud in the center of each plant should sit exactly level with the soil surface.

Gardeners who grow strawberries in containers in a disease-free soil mix don't have to worry about verticillium wilt and red stele (root rot). Both are caused by soilborne fungus. When growing strawberries in containers or in garden soil, ask for plants that are certified as disease free.

Winter protection is needed where alternate freezing and thawing of the soil may cause the plants to heave and break the roots. Low temperatures also injure the crowns of the plants.

Place a straw mulch 3 or 4 inches deep over the plants before the soil is frozen hard. Remove most of the mulch in spring when the centers of a few plants show a yellow-green color. You can leave an inch of loose straw, even add some fresh straw between rows. The plants will come up through it, and it will help retain moisture in the soil and keep mud off the berries.

In northern areas, and as far south as North Carolina, strawberries should be set out in early spring. In the warmer regions of North Carolina, plants can be set out in fall or winter as well, and you can expect a light crop from these plants about five months later.

In northern Cotton Belt climates, set plants in September for the highest yield of spring berries. Waiting until later will diminish the crop.

In the warmest Gulf climates and into Zone 2, and Florida, you must order cold-stored plants, or plants from the north, for planting from February to March. You can also obtain a quick crop by planting northern plants in early November for winter fruiting, but the crop will be smaller. The runners from February plantings can be transplanted in May and August to increase the size of your planting.

PLANTING STRAWBERRIES

Plant strawberries in a double row hill system with furrows on the outside or in a single row with space on either side of the plants for the runners to establish into the matted row system.

Plant strawberries with the roots spread in a fan shape. Keep the crown above the soil line.

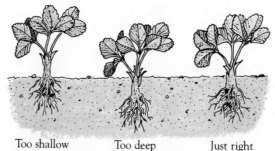

Too shallow   Too deep   Just right

DOUBLE ROW HILL SYSTEM

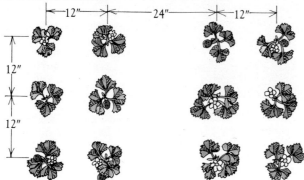

MATTED ROW SYSTEM:

Two ways of spacing the runners

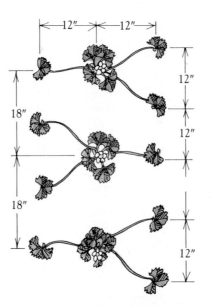

Western Planting Seasons

Western planting seasons impose unique restrictions on strawberry growth. In the coldest areas of Zone 7 (California) and Zones 6 and 7 (Great Basin to Colorado), plant as early in spring as possible since there is good moisture in the soil to start the plants. If soils are usually too wet, protect a mounded bed with plastic to keep it friable.

In the Northwest, especially the coolest areas of western Washington, plant early in fall so plants can become established before real cold sets in, or else wait until early spring. Watch for washouts from heavy fall rain. Weed carefully in spring so weeds don't compete.

In California, plant in fall in cold-winter zones (12, 13, and 9), but get the plants in early in Zone 7. In midwinter zones (10 and 11), use chilled plants (stored at 34°F for a short period) and set them out in October and early November. The low desert is a chancy area for strawberries, but October planting may give results. Everbearing strawberries are now available for California. These everbearing plants are often referred to as "day-neutral" strawberries. The changes in day length through winter and summer do not affect their fruit-bearing capability. They are best planted in the fall and will bear throughout the following year.

Anywhere in California, the berries will do better with a plastic mulch, which increases winter soil temperature and keeps the berries off the soil. Irrigate by furrows for raised beds or by drip irrigation.

Varieties for the South

◀ **'Albritton'.** Origin: North Carolina. This late berry is large and uniform in size, and excellent fresh and for freezing. It develops a rich flavor in North Carolina.

◀ **'Blakemore'.** Origin: Maryland introduction. These early berries are small and firm, with a high acid and pectin content. They have only fair flavor but are excellent for preserves. Plants are vigorous, with good runner production and high resistance to virus diseases and verticillium wilt. They are adapted to a wide range of soil types from Virginia to Georgia, and westward to Oklahoma and southern Missouri.

◀ **'Daybreak'.** Origin: Louisiana. These medium-red berries are large and very attractive, with good flavor and preserving quality. Plants are very productive. Locally available.

◀ **'Dixieland'.** This early berry is deep red, firm, acid in flavor, and excellent for freezing. Plants are sturdy and vigorous.

◀ **'Earlibelle'.** This widely adapted early variety produces large, firm fruit that is good for canning and freezing. Plants are medium size, with good runner production and resistance to leaf spot and leaf scorch.

Left: *This permanent screen keeps strawberries safe from birds and squirrels.*

Below: Midway

Below: 'Sparkle'

◀ **'Florida 90'.** Origin: Florida. Berries are very large, with very good flavor and quality. The plant is a heavy producer of fruit and runners.

◀ **'Guardian'.** These large, deep red, mid-season berries are firm, uniform in size, and attractive. They have good dessert quality and freeze well. Plants are vigorous and productive, and resist many diseases.

◀ **'Headliner'.** Origin: Louisiana. These midseason berries are of good quality. Plants are vigorous and productive, make runners freely, and resist leaf spot. Locally available.

◀ **'Marlate'.** This very large, attractive fruit is good fresh and freezes well. The plant is extremely hardy, and is, therefore, a productive and dependable late variety.

◀ **'Pocahontas'.** Origin: Maryland. This berry is good fresh, frozen, or in preserves. Plants are vigorous and resist leaf scorch. They are adapted from southern New England to Norfolk, Virginia.

◀ **'Redchief'.** The fruit is medium to large and of uniform deep red color with a firm, glossy surface. The plant is extremely productive and resistant to red stele.

◀ **'Sunrise'.** Origin: Maryland. Berries are medium size, symmetrical, and firm, with very good flavor. The flesh is too pale for freezing. A vigorous grower, the plant resists red stele, leaf scorch, and mildew.

◀ **'Surecrop'.** Origin: Maryland. This early berry is large, round, glossy, firm, and of good dessert quality. Large plants should be spaced 6 to 9 inches apart for top production. It resists red stele, verticillium wilt, leaf spot, leaf scorch, and drought. Good in all zones. Widely available.

◀ **'Suwanee'.** This is a medium to large, early, tender berry of very good quality, both fresh or frozen. It is a poor shipper, but excellent for the home garden. Locally available.

◀ **'Tennessee Beauty'.** This late berry is medium size, attractive, glossy red, firm, and has good flavor. It is good for freezing. Plants are productive in fruit and runners. They resist leaf spot, leaf scorch, and virus diseases.

Varieties for the Northeast and Midwest

◀ **'Ardmore'.** Origin: Missouri. These large, late midseason berries are yellowish-red outside, lighter inside, and have good flavor. Plants are productive in heavy silt loam.

◀ **'Catskill'.** These large midseason berries are of good dessert quality and are excellent for freezing. Fruit is not firm enough for distance shipping, but the plant is a productive home garden variety. It can be grown over a wide range of soil types from New England and New Jersey to southern Minnesota. Widely available.

◀ **'Cyclone'.** Origin: Iowa. This variety yields large berries, with very good flavor, that are good for freezing. The plant is hardy, resists foliage diseases, and is well adapted to the North Central states. Widely available.

◀ **'Dunlap'.** Origin: Illinois. This early to midseason fruit is medium size, with dark crimson skin and deep red flesh. It does not ship well but is a good home garden fruit. Plants are hardy and adapted to a wide range of soil types in northern Illinois, Iowa, Wisconsin, Minnesota, North Dakota, South Dakota, and Nebraska.

◀ **'Fletcher'.** Origin: New York. Berries are medium size, with a medium red, glossy, tender skin. Flavor is excellent. They are

also very good for freezing. Plants are well adapted to New York and New England.

◀ **'Howard 17' ('Premier').** Origin: Massachusetts. These are early, medium-size berries of good quality. Plants are productive and resistant to leaf diseases and virus diseases. Locally available in the Northeast.

◀ **'Midland'.** Origin: Maryland. This very early variety bears large, glossy berries with deep red flesh. They are good to excellent fresh and freeze well. The plant does best when grown in the hill system. It is adapted from southern New England to Virginia, and west to Iowa and Kansas.

◀ **'Midway'.** Origin: Maryland. These large berries are of good to very good dessert quality and are good for freezing. Plants are susceptible to leaf spot, leaf scorch, and verticillium wilt. They are widely planted in Michigan. Widely available.

◀ **'Raritan'.** Origin: New Jersey. These midseason berries are large, firm, and have good flavor. Plants are medium size.

◀ **'Redstar'.** Origin: Maryland. These late berries are large and of good to very good dessert quality. Plants resist virus diseases, leaf spot, and leaf scorch. They are grown from southern New England, south to Maryland, and west to Missouri and Iowa.

◀ **'Sparkle'.** This is a productive mid-season variety, with bright red, attractive berries that are fairly soft and have good flavor. Berry size is good in early picking, but small in later ones. Widely available.

◀ **'Trumpeter'.** Origin: Minnesota. These medium-size late berries are soft and glossy and have very good flavor. This is a hardy and productive home garden variety for the upper Mississippi Valley and the Plains states.

Left: 'Olympus'

Below: 'Sequoia'

Everbearing Varieties
◄ **'Gem' ('Superfection'** and **'Brilliant'** are considered to be nearly identical to 'Gem'). This variety yields small, glossy red, tart fruit of good dessert quality. Widely available.

◄ **'Geneva'.** Large, vigorous plants fruit well in June and throughout the summer and early autumn. Berries are soft and highly flavored.

◄ **'Ogallala'.** Berries are dark red, soft, and of medium size; have good flavor; and are good for freezing. Plants are vigorous and hardy. Widely available.

◄ **'Ozark Beauty'.** An everbearing variety for the cooler climate zones, this one produces poorly in mild climates. The berries are bright red inside and out, are large, sweet, and of good flavor. In any one season, only the mother plants produce, yielding crops in the summer and fall. Runner plants produce the following season. Widely available.

Varieties for the West
The western strawberry growing regions are divided into three areas: The Rockies and the Great Basin, western Washington and Oregon, and California.

The Rockies and the Great Basin
Recommended for these areas are these varieties, described earlier: 'Cyclone', 'Dunlap', 'Gem', 'Ogallala', 'Ozark Beauty', 'Sparkle', and 'Trumpeter'.

Western Washington and Oregon
◄ **'Hood'.** Origin: Oregon. These midseason berries are large, conical, bright red, and glossy. They are held high in strong, upright clusters, and are good fresh or in preserves. Plants are resistant to mildew but somewhat susceptible to red stele.

◄ **'Northwest'.** Origin: Washington. Late midseason fruit is large at first, then smaller, with crimson skin and red flesh. It is firm, well flavored, and good fresh, in preserves, or for freezing. Plants are very productive and so resistant to virus diseases that they can be planted where virus has killed other varieties.

◄ **'Olympus'.** Origin: Washington. Late midseason fruit is held well up on arching stems. Berries are medium to large, bright red throughout, tender, and firm. Plants are vigorous but produce few runners. They resist red stele and virus diseases but are somewhat susceptible to botrytis infection.

◄ **'Puget Beauty'.** Origin: Washington. The large, glossy, very attractive fruit has light crimson skin. Flesh is highly flavored, excellent fresh, and good for freezing and preserves. Plants are large and upright, with moderate runner production. They resist mildew but are somewhat susceptible to red stele. Locally available.

◄ **'Quinault'.** An everbearer with a moderate early crop, heavier in July–September. The fruit is large and soft, with good color. The plant produces good runners.

◄ **'Rainier'.** These late midseason berries are large, firm, and of good quality. Plants are vigorous, with large leaf blades but moderate runner production.

◄ **'Shuksan'.** This midseason variety bears large, firm berries that are bright red, glossy, and broadly wedge shaped. Fruit is good for freezing, and plants are vigorous.

California
◄ **'Douglas'.** Origin: California. Large, uniform, midseason fruit, which is light red and firm. The plant is very vigorous and produces early berries when planted in October. Good in Southern California.

◄ **'Pajaro'.** Origin: California. Large, uniform, early fruit with red skin and flesh. This berry does well along the central coast for spring and summer berries. It does well in the Central Valley for early season berries. Locally available.

◄ **'Sequoia'.** Origin: California. This is an early variety that may even bear in December. The exceptionally large fruit is dark red and tender with soft flesh and has excellent flavor. Harvest frequently for best quality. The plant is erect and vigorous, with many runners. It is recommended for home gardens in the central and south coast. Plant in October and November. Widely available.

◄ **'Shasta'.** Origin: California. This large midseason berry is bright red and glossy with firm red flesh, good for freezing or preserves. Plants are fairly vigorous, with a moderate number of runners. They have some resistance to mildew and virus diseases. Locally available.

◄ **'Tioga'.** Origin: California. This early berry is medium red and glossy, with firm flesh that is fine for preserves or freezing. The plant is vigorous, moderately resistant to virus, and fairly tolerant of salinity, but highly susceptible to verticillium wilt. It is good for late-summer planting.

◄ **'Tufts'.** Origin: California. This midseason berry is red, extremely firm, and very large. This is a good variety for Southern California.

Everbearing Varieties for California
These are heavy berry producers and do not form many runners. Plant them in the fall for fruit production all through the next year. They produce medium-size berries. Try 'Aptos', 'Brighton', and 'Hecker'.

Select fruit trees from your local nursery, where you can pick them out yourself. If your nursery doesn't stock the varieties you want, order from the list below.

CATALOG SOURCES

The following descriptions will tell you which companies are specialists in fruits, and which cover a large spectrum of garden materials. Many of these catalogs are excellent reference sources for the careful reader.

If not marked otherwise, sources are retail. Wholesale nurseries will not sell directly to you, but you can refer your local nursery to them as sources for plants you wish to buy.

Adams County Nursery & Fruit Farms
Aspers, PA 17304
Fruit specialists. Wholesale and retail.

W. F. Allen Co.
P.O. Box 1577 H
Salisbury, MD 21801
Strawberry specialists. Catalog and planting guide lists over 30 varieties. Wholesale and retail.

Armstrong Nurseries
P.O. Box 4060
Ontario, CA 91761
Specialists in fruit trees and roses. Catalog lists genetic dwarf peaches and nectarines, as well as other exotic fruit trees. Vegetables and bulbs.

Bountiful Ridge Nurseries, Inc.
Princess Anne, MD 21853
Specialists in fruits and nuts. Catalog includes planting guide. Wholesale and retail.

Bowers Berry Nursery
94959 Hwy 99E
Junction City, OR 97448
Catalog of berries and grapes. Wholesale and retail.

Bryants Nursery
P.O. Box 422
Helena, GA 31037

Buckley Nursery Co.
Buckley, WA 98321
Catalog of fruit, flowers, and shade and ornamental trees. Wholesale and retail.

Bunting's Berries
Selbyville, DE 19975
Fruit trees, berries, and nursery stock.

Burgess Seed and Plant Co.
P.O. Box 82
Galesburg, MI 49053
*Two catalogs are offered:
General seed catalog. Flowers, vegetables, fruit, and nursery stock catalog. Especially for the limited-space gardener. Full color catalog, with a page on fruits to grow indoors.*

W. Atlee Burpee Co.
Warminster, PA 18974
Clinton, IA 53732
Riverside, CA 92502
General seed catalog. Flowers, vegetables, fruit, garden aids, and nursery stock.

C & O Nursery
P.O. Box 116
1700 N. Wenatchee Ave.
Wenatchee, WA 98801
Fruit specialists, exclusive patented varieties. Catalog includes ornamentals and shade trees. Wholesale and retail.

The Clyde Nursery
Highway U.S. 20
Clyde, OH 43410
Catalog of fruits and berries.

Columbia Basin Nursery
Box 458
Quincy, WA 98848
Colored brochure and price list. Seedling rootstock, dwarfing apple rootstock, dwarf and standard budded fruit trees. Wholesale and retail.

Cumberland Valley Nurseries, Inc.
P.O. Box 430
113 Lind Street
McMinnville, TN 37110
Catalog specializing in plums, peaches, and nectarines. Wholesale and retail.

Farmer Seed & Nursery Co.
Fairbault, MN 55021
General seed catalog. Flowers, vegetables, fruit, and nursery stock.

Henry Field Seed & Nursery Co.
407 Sycamore Street
Shenandoah, IA 51601
General seed and nursery catalog. Flowers, vegetables, fruit, gardening aids, and nursery stock.

Dean Foster Nurseries
Hartford, MI 49257
General catalog specializing in strawberries. Flowers, vegetables, dwarf fruit, and berries. Wholesale and retail.

Four Winds Growers
P.O. Box 616
Mission San Jose, CA 94538
Dwarfed citrus trees a specialty. Instructional booklet available to the home gardener.

Fowler Nurseries, Inc.
525 Fowler Road
Newcastle, CA 95658
Price list of over 200 varieties sent on request. Commercial price list also available. Catalog $1.

Grootendorst Nurseries
Lakeside, MI 49116
Specialists in dwarf Malling and Merton rootstock.

Gurney Seed & Nursery Co.
1448 Page St.
Yankton, SD 57078
General seed catalog. Flowers, vegetables, fruit, and nursery stock.

Haley Nursery Co., Inc.
Smithville, TN 37116
Price list on fruit trees, specializing in peaches and nectarines. Wholesale only. Ask your dealer to order.

H. G. Hastings Co.
Box 4655
Atlanta, GA 30302
General seed catalog. Flowers, vegetables, fruits, and nursery stock.

Heath's Nursery, Inc.
P.O. Box 707
Brewster, WA 98812
Catalog of fruit, shade, and ornamental trees.

Hilltop Orchards & Nurseries, Inc.
Rt. 2
Hartford, MI 49057
Widely recognized fruit tree specialists for commercial orchardists. Free handbook and catalog.

Inter-State Nurseries
Hamburg, IA 51640
Fruit, flowers, berries, roses, and ornamentals.

Ison's Nursery & Vineyard
Brooks, GA 30205
Catalog specializing in grapes. Wholesale and retail.

J. W. Jung Seed Co.
Station 8
Randolph, WI 53956
General seed catalog. Flowers, vegetables, fruit, and nursery stock.

Kelly Bros. Nurseries, Inc.
Dansville, NY 14437
Catalog of fruit, nuts, flowers, and ornamentals.

Lakeshore Tree Farms, Ltd.
R.R. 3
Saskatoon, Sask. S7K 3J6
Canada

Lawson's Nursery
Route 1, Box 294
Ball Ground, GA 30107
Fruit catalog specializing in old-fashioned and unusual fruit trees. Lists over 100 varieties of old apples.

Henry Leuthart Nurseries, Inc.
East Moriches, Long Island, NY 11940
Fruit catalog and guidebook on dwarf and espaliered fruit trees.

Earl May Seed & Nursery Co.
Shenandoah, IA 51603
General seed catalog. Flowers, vegetables, fruit, and nursery stock.

Mayo Nurseries
Route 14
Lyons, NY 14489
Fruit specialists. Catalog includes many varieties of dwarf and semidwarf apples. Wholesale and retail.

Miller's Nursery, Inc.
Canandaigua, NY 14424
Fruit specialists. Catalog also includes garden aids and ornamentals.

New York State Fruit Testing
Cooperative Association
Geneva, NY 14456
Fruit catalog. $4 membership fee, refunded on first order.

L. L. Olds Seed Co.
P.O. Box 7790
Madison, WI 53701
General seed catalog. Flowers, vegetables, fruit, and nursery stock.

Owen's Vineyard and Nursery
Georgia Highway 85
Gay, GA 30218
Catalog specializing in muscadine grapes. Includes guidelines for growing and training. Southern rabbiteye blueberries are available.

Ponto Nursery
656 West Quarry Road
San Marcos, CA 92069
Specializing in citrus: wholesale only. No direct sales or catalog.

Preservation Apple Tree Company
First and Birch Streets
Mt. Gretna, PA 17064
Wholesale and mail order.

Rayner's Bros., Inc.
P.O. Box 1617
Salisbury, MD 21801
Fruit catalog specializing in strawberries.

Southmeadow Fruit Gardens
2363 Tilbury Place
Birmingham, MI 48009
Probably the largest collection of fruit varieties, old, new, and rare, in the United States. Illustrated catalog is priced at $5 and worth it. A condensed catalog is free.

Stanek's Garden Center
East 2929 — 27th Avenue
Spokane, WA 99203
Catalog of fruit, berries, flowers, and ornamentals.

Stark Bros. Nursery
Louisiana, MO 63353
Illustrated catalog and guide to vegetables, fruit, nuts, and ornamentals.

Van Well Nursery
P.O. Box 1339
Wenatchee, WA 98801
Fruit and berry catalog. Wholesale and retail.

Waynesboro Nurseries
P.O. Box 987
Waynesboro, VA 22980
Catalog of fruits, nuts, and ornamental plants.

Weeks Berry Nursery
6494 Windsor Island Road, North
Salem, OR 97303
Specializing in small fruits. Wholesale and commercial plantings.

Dave Wilson Nursery
Box 90 A
Hughson, CA 95326
Fruits, berries, grapes. Specializing in Zaiger patented fruit trees. Wholesale and retail.

INDEX

Botanical names have been listed and cross-referenced to the common name. Numbers in italics refer to pages with illustrations.